Propagation and Genetic Manipulation of Plants

Iram Siddique
Editor

Propagation and Genetic Manipulation of Plants

Springer

Editor
Iram Siddique
Department of Botany
Aligarh Muslim University
Aligarh, Uttar Pradesh, India

ISBN 978-981-15-7735-2 ISBN 978-981-15-7736-9 (eBook)
https://doi.org/10.1007/978-981-15-7736-9

This Springer imprint is published by the registered company Springer Nature Singapore Pte Ltd.
The registered company address is: 152 Beach Road, #21-01/04 Gateway East, Singapore 189721, Singapore

Contents

1 **Somaclonal Variation and Methods Used for Its Detection** 1
V. M. Ranghoo-Sanmukhiya

2 **Synthetic Seed Technology in Some Ornamental and Medicinal Plants: An Overview** 19
Mehpara Maqsood, Mir Khusrau, A. Mujib, and Zahoor A. Kaloo

3 **Cytokinin Influence on Micropropagation System of *Dianthus caryophyllus* L.** 33
Iram Siddique and Vikas Yadav

4 **Phenolic Compound Contents of *Hypericum* Species from Turkey** 43
Hilal Surmuş Asan

5 **Morphogenesis of Cotyledon Explants and Plantlet Regeneration of Tomato (*Lycopersicon esculentum* Mill.) Under Salt Stress** 69
M. M. A. N. Shanika and Thayamini H. Seran

6 **Callus Culture Systems in *Salix* L.: The Limited Database** 83
Afsheen Shahid and Mohammad Anis

7 **An Efficient Plant Regeneration System Via Leaf Derived Callus of *Solanum melongena* L.** 93
Hina Khan

8 **Selection of Elite Genotype of a Multipurpose Forest Tree (*Pterocarpus marsupium* Roxb.) from Naturally Grown Populations in Central India** 101
Anees Ahmad, Iram Siddique, and Ankita Varshney

9 **Regulation of In Vitro Morphogenesis by Modulation of Culture Conditions in *Withania somnifera* L. Using Cotyledonary Node Explants** 121
Nigar Fatima and M. Anis

10 Substitution of Benzyladenine with *meta*-Topolin Improved Shoot Regeneration and Rooting in *Wedelia chinensis* (Osbeck) Merr. and Quantitative Estimation of Flavonoids 139
Mehrun Nisha Khanam, Saad Bin Javed, Naseem Ahmad, and Neetu Singh

11 Effect of Silicon Ions on In Vitro Morphogenesis in *Rauvolfia tetraphylla* Using Shoot Tip Explants 153
Sheikh Altaf Hussain

12 TDZ-Induced Efficient Micropropagation from Juvenile Nodal Segment of *Syzygium cumini* (Skill): A Recalcitrant Tree 163
Afshan Naaz, Iram Siddique, and Altaf Ahmad

About the Editor

Iram Siddique is an Assistant Professor in the Department of Botany, Aligarh Muslim University, Aligarh, India. She has also worked as an Assistant Professor in the Department of Botany and Microbiology, King Saud University, Riyadh, Saudi Arabia, where she was also a Co-Principal Investigator in a research group project funded by DSR-KSU, Riyadh. She has also worked as a Senior Research Fellow and Research Associate of CSIR, New Delhi. She has more than ten years of teaching and research experience in Plant Tissue Culture and Molecular Biology. She has published about 35 research papers in the journals of International repute. The same have been cited 446 times with 11 h-index. She has also participated in several national and international symposiums and seminars held at various universities and institutes of India and abroad. She was conferred by a number of prestigious awards like Young Scientist award of the Year 2010 of UPCST, Lucknow, Junior Scientist of the Year 2011 of NESA, New Delhi, and Excellence in Research (Scientist) award 2018. She visited Faculty of Agriculture, University of Mauritius, Mauritius, Singapore, Addis Ababa University, Ethiopia, and UAE under various research exchange programs. She holds membership of national and International scientific bodies and editorial boards of various journals.

Somaclonal Variation and Methods Used for Its Detection

1

V. M. Ranghoo-Sanmukhiya

Abstract

During plant tissue culture, the genetic stability of in vitro plantlets is often compromised and the underlying factors include the plant species or varieties involved, mode of regeneration used, genotype and ploidy level, composition of the growth medium, duration of the callus phase, and total time in culture. During tissue culture, conversion of the explant into in vitro tissue in fact imposes stress to the plant cells, which undergo a genomic shock and require restructuring of their genomes. An array of morphological, biochemical, and molecular methods has been used to assess somaclonal variation in different plant species and variable efficiency has been observed. The array of methods including molecular markers used for ensuring genetic stability of regenerated cultures in different plants is reviewed in this chapter.

Keywords

Somaclonal variation · Tissue culture · Molecular markers

1.1 Introduction

Tissue culture experiments involve growing plant tissues under a set of specific laboratory conditions different from their normal growth conditions in nature in an aim to eventually propagate and conserve them or for ultimate genetic transformation. When the explant is placed in vitro, a specific tissue culture protocol has to be optimized or designed for each species used. Moreover, adjustments in physiology, anatomy, and metabolism have to be made for them to be able to grow and

V. M. Ranghoo-Sanmukhiya (✉)
Faculty of Agriculture, University of Mauritius, Réduit, Moka, Mauritius
e-mail: m.sanmukhiya@uom.ac.mu

I. Siddique (ed.), *Propagation and Genetic Manipulation of Plants*,
https://doi.org/10.1007/978-981-15-7736-9_1

multiply in culture. During tissue culture, any plant cell type or organ including embryos, spores, roots, leaves, and protoplast are used as explants. There is normally a cycle of dedifferentiation of tissues from the explant referred to as (callus production) under conditions set by the researcher followed by proliferation for a number of generations (callus proliferation) culminating in the regeneration of the plants (Larkin and Scowcroft 1981). Regeneration only involves repeated mitotic divisions, which ideally should yield genetically uniform plants (Larkin 1998). The "somaclone" is the regenerated plant obtained following tissue culture, while "calliclones" and "protoclones" refer to regenerated plants obtained from stems and protoplast (Chawla 2000).

During tissue culture, conservation of the genetic integrity of the mother plant is a desirable component, but regeneration by tissue culture has been reported to yield plants which exhibit unexpected and undesired phenotypic anomalies and variation (Karp 1994). Conditions inherent in vitro include regimented media components, low light levels, and more importantly high humidity which eventually cause several physiological and developmental aberrations of the regenerated plantlets. In fact, somaclonal variation refers to culture-induced anomalies or variation, which eventually becomes genetic and is inherited by the clonal progenies (Chawla 2000; Us-Camas et al. 2014). There are other variations induced in culture which are only temporary and reverse back to the normal state when faced with other culture conditions. The temporary changes which are nonheritable and not reversible are in fact caused by epigenetic and physiological factors, otherwise heritable and permanent changes represent "de novo" genomic variation (Kaeppler et al. 2000). Initially, somaclonal variants were classified as artifacts caused by overexposure to phytohormones and epigenetic events (Chawla 2000; Isah 2015). In fact, genetic variation observed following regeneration is a result of aberration of genome expression which is due to the formation of chromosome mosaics and spontaneous mutation. Moreover, somaclonal variation can be induced by a directed approach through in vitro selection and can be minimized as well without any in vitro selection (Brar and Jain 1998; Chawla 2000). The genetic basis of somaclonal variation, its uses in agriculture, associated disadvantages and methods of detection are discussed.

1.2 Uses of Somaclonal Variation in Agriculture

In the current era, we need to secure food availability faced with the impending crisis of global warming and declining arable soil. Somatic embryogenesis can be exploited to produce plants with certain variable characteristics as compared to the starting material and without using transgenics. Under tissue culture conditions, the plant will do what it is programmed to do but variation will result by using epigenetic regulation and hence inducing variation. In addition, the number of embryos produced can be increased by making use of specific chemicals, e.g., 5-Aza for plants which are recalcitrant for SE and hence increase production. Somaclonal variation can be considered as an advantage as it allows creation of new genotypes which have been exploited in agriculture for the following:

1.2.1 Generation of New Agronomic Variants with Favorable Traits

Somaclonal variation has given rise to new breeding lines with variants of important agronomic traits. Many variants in different crops have been generated using somaclonal variation which can further be integrated in breeding programs for introgression of important agronomic traits in different lines. Somaclonal variants were selected for: high oil yield in *Cymbopogon winterianus* (Dey et al. 2015), higher flowering and fruiting ability in strawberry (Biswas et al. 2009), high sucrose yielding sugarcane varieties (Rastogi et al. 2015), finger millet of higher seed and biomass yield (Baer et al. 2007), non-browning potato varieties (Arihara et al. 1995; Thieme and Griess 2005).

1.2.2 Generation of Disease Resistance Varieties

Disease resistance for different crop varieties has been obtained through induced somaclonal variation. Somaclonal variants from different crops with increase disease resistance include sugarcane which were resistant to eyespot and red rot disease (Rastogi et al. 2015), wheat resistant to white blotch disease (Arun et al. 2003), and *Fusarium* wilt-resistant banana (Asif and Othman 2005).

1.2.3 Generation of Biochemical Variants with Abiotic Resistance

Biochemical variants can be generated and these can be used in plant metabolic pathway studies. Somaclonal variants in wheat which exhibit frost and freezing tolerance have been obtained and these have been directly related to the production of proline (Dorffling and Melz 1997). Somaclonal variation has also successfully yielded plants which exhibit salt and herbicide resistance in various crop species, e.g., drought and salt resistance in sugarcane (Rastogi et al. 2015) as well as drought tolerance rice (Adkins et al. 1995).

1.3 Disadvantages of Somaclonal Variation

1. Somaclonal variation generates uncontrollable and unpredictable variation which are of no agricultural use. Most changes are cultivar independent.
2. Genetic changes during regeneration occur at variable frequencies and are not stable and nonheritable.
3. Somaclonal variants are not novel genotypes and have frequently to be checked for novelty.
4. Generation of somaclonal variants is limited only to those plants which have the power to multiply in culture and regenerate into whole plants.
5. Somaclones have reduced growth rate, reduced fertility as well as low-performance rate. They sometimes lose the power of regenerating.

1.4 Causes and Genetic Basis of Somaclonal Variation

Conditions in tissue culture which induce mutations include wounding, removal of the cell wall, sterilization which includes exposure to high concentration of chemicals including sterilants, sugars, and plant growth regulators in the nutritive media. Genetic alteration of the regenerants in culture is inevitable and increases with prolongations (Duncan 1996; Us-camas et al. 2014). The in vitro environment is stressful and hampers proper morphogenesis of tissues and this is due to erratic gene expression which results in disrupted hormonal synthesis and signaling. In addition, oxidative stress has been reported in plant tissue culture which normally results in elevated reactive oxygen species production (Isah 2015). During regeneration of plants using tissue culture, there is normally differentiation and redifferentiation of cells leading to qualitative and quantitative genomic changes. Somaclonal variation can be traced back to changes which have occurred at the genome level and include:

1. Single gene mutation: Somaclonal variants in tomato have been linked to single gene mutation inside the genome of crop species including wheat, maize, rice (Edallo et al. 1981), and tomato (Evans and Sharp 1983).
2. Activation of transposable elements: Transposable elements have been linked with somaclonal variation and were first reported to cause colored spots in *Nicotiana tabacum* species (Lorz and Scowcroft 1983). Activation of the transposable elements has been observed in in vitro maize and tobacco tissues (Peschke and Phillips 1991). In vitro genetic rearrangements (Hirochika et al. 1996; Sato et al. 2011) as well as insertion and excision of transposable elements have also been observed (Gupta 1998; Sato et al. 2011).
3. Karyotype changes: Variant plants with altered chromosome numbers inducing ploidy level changes namely polyploidy which includes chromosomal abnormality have been observed (Hang and Bregitzer 1993; Mujib et al. 2007; Leva et al. 2012).
4. Changes in chromosome structure: Changes in chromosome structure which include chromosome breakage and rearrangement, centric and acentric fragments, abnormal arrangement of chromosomes, ring chromosomes, and formation of micronuclei (Kaeppler et al. 1998; Alvarez et al. 2010).
5. Plastid genetic change: Plastid genetic changes including deletion, addition, and intramolecular recombinations have occurred in somaclones (Cassells and Curry 2001; Bartoszewski et al. 2007).
6. Methylation: Methylation in single-copy sequences, methylation of the genome as well as histone modifications have been reported in tissue culture (Kaeppler et al. 2000; Linacero et al. 2011; Miguel and Marum 2011).

1.5 Methods Used for Assessing Somaclonal Variation

1.5.1 Morphological Assessment of Somaclonal Variation

Morphological assessment of somaclonal variation has been traditionally used to detect "variants" been referred to as "off types" regenerated plantlets. Characters most often used for the assessment include plant stature, leaf morphology, and pigmentation. Examples of somaclonal variants include banana dwarf off-types (Rodrigues et al. 1998) or excessive vegetative growth in palms (Zaid and Al Kaabi 2003). Morphological assessment for somaclonal variation may not be reliable since the number of characters used may be limited and more importantly be artifacts and caused by the impeding stress inherent of the tissue culture conditions (Cloutier and Landry 1994). Moreover, genomic changes caused during tissue culture but these may not be reflected in the phenotype (Bairu et al. 2011).

1.5.2 Biochemicals Used to Assess Somaclonal Variation

Responses of explants to certain chemicals such as gibberellic acid have been used for the assessment of somaclonal variation (Phinney 1985; Sandoval et al. 1995; Graebe 2003). Gibberellic acid is well known for its regulation of growth and development and somaclonal variants differ from normal plantlets in having disturbed gibberellic acid metabolism and endogenous level allowing somaclone identification (Bairu et al. 2011). Normal banana plants have been found to be gibberellic acid responsive and hence developed significantly greater leaf sheaths as compared to dwarf somaclonal variants which were nonresponsive (Damasco et al. 1996). Synthesis of chlorophyll and carotenoids has been used to differentiate somaclones with normal plantlets (Mujib et al. 2007; Wang et al. 2007).

1.5.3 Protein Markers Used to Assess Somaclonal Variation

Isozymes are different molecular forms of proteins which are used to actively control biochemical processes inside cells (Kunert et al. 2003). Isozymes fingerprints are normally generated and can be used as a basis for finding differences between genomes of parental plants and regenerated individuals. Enzymes commonly measured include esterase (EST), endopeptidase, alcohol dehydrogenase, peroxidase, polyphenol oxidase amongst others (Kunert et al. 2003). Isozyme analysis has advantages such as facilities of usage and low cost. Isozymes assays, however, have limited polymorphisms which are highly influenced by environmental conditions (Kunert et al. 2003). Somaclonal variation has been assayed using isozymes in palms (Salman et al. 1988; Saker et al. 2000), soybean (Amberger et al. 1992), citrus (Carini and De Pasquale 2003), and potato (Afrasiab and Iqbal 2012).

1.5.4 Cytogenetic Analysis for the Determination of Somaclonal Variation

Chromosomal abnormalities can occur in tissue culture plantlets and could be the causal factors of somaclonal variations (D'Amato 1985; Kaeppler et al. 2000). Presence of an abnormal number of chromosomes, aneuploidy, has been reported during callus induction and cell suspension phases of citrus and oil palm (Hao and Deng 2002; Giorgetti et al. 2011). Chromosomal alterations due to non-separation of chromatids during mitosis and meiosis can lead to the production of chromosomal abnormalities and segmental or complete aneuploidy (D'Amato 1985; Holland and Cleveland 2009; Kaeppler and Phillips 1993; Kaeppler et al. 2000). Variations in chromosome number among cells in cultures can be detected using fluorescence in situ hybridization, flow cytometry, and microscopic examination. Flow cytometry is a technique which allows the calculation of the DNA content analysis through the use of a fluorochrome specific to the DNA found in whole cells, protoplast, nuclei, or chromosomes. Using flow cytometry, diploids and tetraploids were recovered from calli cells of *Arapidopsis* which had cytogenetic abnormalities such as chromosomal translocations, deletions, and duplications (Orzechowska et al. 2013). Flow cytometry has allowed detection of chromosomal abnormalities during somaclonal variation in many plants including eucalyptus (Pinto et al. 2004), cotton (Jin et al. 2008), papaya (Abreu et al. 2014), and *Viola uglinosa* (Slazak et al. 2015).

1.5.5 Detection of Somaclonal Variation Using RFLP

The restriction fragment length polymorphism (RFLP) is one technique which is used for genome analysis and it involves digestion of genomic DNA using enzymes restriction endonucleases, followed by separation of the digested products on agarose gel for the generation of DNA profiles. Variations in the number and size of the bands will occur in the DNA profiles of the true to type regenerated plantlet as compared to the somaclonal variant. Variations in the band number and sizes observed inside the genome can be explained by insertion, deletion, or single nucleotide polymorphisms (Agarwal et al. 2008; Bairu et al. 2011). RFLP markers have been used to screen for somaclonal variation in rice (Müller et al. 1990), maize (Brown et al. 1991) sugar beet (Levall et al. 1994), *Hypericum perforatum* (Halušková and Čellárová 1997), and oil palm (Jaligot et al. 2002).

1.5.6 Detection of Somaclonal Variation Using Random Amplified Polymorphic DNA (RAPD)

Random amplified polymorphic DNA (RAPD) is a PCR-based technique which relies on the selective amplification of fragments by using primers of a random nature. These primers are around ten nucleotides in length which have the ability to bind to complementary sites within the genome followed by amplification of the sites delimited. RAPD represents one among the most popular markers used for the assessment of somaclonal variation due to low input cost and low technology required. Presence of certain bands in the parental strains and absence in regenerated plantlets indicates genomic rearrangement during tissue culture and account for somaclonal variation. Common bands which are found in the somaclones originate from mutation inherited from the mother embryogenic callus tissues. Variations in the genome level caused by somaclonal variation include nucleotide substitution events such as transition and transversion (Ngezahayo et al. 2007). RAPDs have been however associated with low reproducibility, reliability, and transferability among different labs. RAPDs have actually been used in the assessment of somaclonal variation in sugarcane (Taylor et al. 1995; Tawar et al. 2008), banana (Trujillo and Garcia 1996; Tang 2005; Abdellatif et al. 2012), orchids (Chen et al. 1998), rice (Yang et al. 1999), maize (Osipova et al. 2001), tomato (Soniya et al. 2001), family of Zingiberaceae (Islam et al. 2004; Wondyifraw and Wannakrairoj 2004; Mohanty et al. 2011), and *Cymbopogon* (Dey et al. 2015).

1.5.7 Detection of Somaclonal Variation Using Simple Sequence Repeat (SSR)

Simple sequence repeat (SSR) also referred to as microsatellite markers consist of short DNA sequence motifs of few base pairs which are repeated tandemly inside the eukaryotic genome. Each genome varies in having a characteristic number of repeats and these can be amplified by using the polymerase chain reaction (PCR) using special primers. Different somaclones will have different numbers of these repeats units which are exhibited with high levels of polymorphism following amplification (Levinson and Gutman 1987; Coggins and O'Prey 1989). SSRs are highly reproducible and codominant which make them ideal, rapid, simple genotyping assays. SSRs are very versatile in asserting genetic variation and are highly used in marker assisted breeding. Moreover, SSR markers have also been used to screen for somaclonal variants in cotton populous (Rajora and Rahman 2003), bananas (Hautea et al. 2004; Ray et al. 2006), rice (Khai and Lang 2005), pine (Marum et al. 2009), sorghum (Zhang et al. 2010), and grapevine (Nookaraju and Agrawal 2012).

1.5.8 Detection of Somaclonal Variation Using the Inter-Simple Sequence Repeat (ISSR) Markers

Inter-simple sequence repeat markers represent the DNA fragments which are found between simple sequence repeats which are oppositely oriented. The ISSRs can be amplified by using primers which are complementary to the microsatellite regions. These inter-simple sequence repeats can be amplified by using the PCR technique which makes use of primers specially designed to amplify part of the microsatellite DNA core sequences as well as selectively amplifying the adjacent regions. The targeted region inside the genomes is found between two closely spaced SSRs which may also be oppositely oriented. Following amplification around 20–50 fragments will be amplified from multiple loci containing the SSR motifs but they will be polymorphic in their sizes. The ISSR technique has been successfully used on several occasions for the genome screening of somaclones of mulberry (Vijayan and Chatterjee 2003), tea (Thomas et al. 2006), and gerbera (Bhatia et al. 2009).

1.5.9 Detection of Somaclonal Variation Using the AFLP Technique

Molecular markers mentioned above targeted highly repetitive sequences inside the genome and genotyping relies on the variation in the number of repeat counts. However, since multiple targets are being exponentially amplified at the same time, DNA profiles obtained may not be reproducible and be variable due to the quality of DNA and other reaction conditions.

The AFLP method involves both restriction enzymes digestion with PCR amplification of fragments obtained and allows generation of fragment length polymorphisms. The AFLP technique relies on the use of primers which are complementary to known adapter sequences to selectively amplify restriction-digested fragments and DNA profiles obtained are more reproducible. The AFLP technique is a more suitable and sensitive marker system with higher reproducibility (Meudt and Clarke 2007). The AFLP technique has been recognized as the ideal marker system for assessing somaclonal variation in *Arabidopsis thaliana* (Polanco and Ruiz 2002), *Coffeea arabica* (Sanchez-Teyer et al. 2003), bean (Rosales-Serna et al. 2005), *Asparagus officinalis* (Pontaroli and Camadro 2005), pea (Smýkal et al. 2007), grapevine (Schellenbaum et al. 2008), and *Feesia hybrida* (Gao et al. 2010).

1.5.10 Detection of DNA Methylation Events Using the Methylation-Sensitive Amplification Polymorphism (MSAP)

DNA methylation of the genome is normally associated with several inheritance processes such as transcriptional silencing, genetic imprinting, and genomic stability. Methyl groups are added to DNA cytosine to form 5-methyl cytosine (He et al. 2011) during the methylation process without any change occurring in the DNA sequence but results in an increase in the compression level of chromatin affecting gene expression eventually as access to transcription by RNA Polymerase II is debarred. The tissue culture environment induces abiotic stress during regeneration and is responsible for global genome-wide methylation. The long callus generation phase allows for the methylation event and induces variation during plant regeneration. In cassava, genome-wide methylation changes have been observed during meristem micropropagation using genome sequencing (Kitimu et al. 2015). DNA methylation events appear to be exclusive to repetitive DNA more specifically to transposons (Lippman et al. 2004). Decrease in the level of methylation may be due to gene activation during cell reprogramming to facilitate plant regenerations.

Changes in methylation have been reported to be the cause of somaclonal variation in many crops (Gonzalez et al. 2013). Epigenetic changes to DNA of plant tissues in culture appear to allow the reactivation of multiple transposon classes. Insertions of the retrotransposon family Tos and that of the class II TE family (nDiaZ) have been reported to increase with tissue culture period (Huang et al. 2009a). Transcriptional activity of the MERE1 retrotransposon family (associated with a decrease in cytosine methylation) has also been found to increase in two accessions of *Medicago truncatula* and has led to 40% of novel insertions mostly within coding sequences (Rakocevic et al. 2009). In grapevines, an increase in TE insertions has also been noted (Lizamore 2013).

The methylation-sensitive amplification polymorphism (MSAP) technique is used for the evaluation of the total methylation level of a genome. The MSAP is actually a modification of the AFLP (Vos et al. 1995) technique in which methylation-sensitive restriction endonucleases such as isoschizomers HpaII and Msp I to digest the DNA. These enzymes have different sensitivity to methylated cytosines and generate digestion patterns which allow identification of the methylated DNA (Yaish et al. 2014). The MSAP technique has efficiently been used to detect somaclonal variants in different plants including coffee and oil palm (Matthes et al. 2001; Bobadilla Landey et al. 2013; Francischini et al. 2017).

1.5.11 Detection of Somaclonal Variation Using Transposon-Based Marker Systems

Transposable elements represent DNA fragments which have the ability to make copies of themselves and can insert into new loci of the genome during transposition. When transposable elements insert themselves within genes there will be a hindrance to normal gene function (Makarevitch et al. 2015) and gene expression will be affected (Hollister and Gaut 2009; Lisch 2009). Induction of transposable elements has been observed in rice (Huang et al. 2009b) and maize (Makarevitch et al. 2015). Retrotransposons were also identified in tobacco and rice and have been reported to be involved with induced mutations in tissue culture (Hirochika 1993; Hirochika et al. 1996).

With the advent of genome sequencing projects, transposable elements have been recognized as being dynamic and abundant for the development of markers namely sequence-specific amplified polymorphism (SSAP), inter-retrotransposon-amplified polymorphism (IRAP), and retrotransposon-based insertion polymorphisms (RBIP) (Fig. 1.1). The sequence-specific amplified polymorphism (SSAP) marker which made use of transposon-specific sequence within an AFLP reaction was developed by Waugh et al. (1997). The SSAP technique has been found to be more specific than SSR-based markers in different agricultural crops namely oats, grapevine, tomato, sweet potato, and pea (Berenyi et al. 2002; Labra et al. 2004; Tam et al. 2005; Lizamore 2013). IRAP markers make use of the long terminal repeat primers (singly or in pairs) to amplify different retrotransposons distributed inside the genome of the somaclones, hence, amplifying regions of the genome which are found between

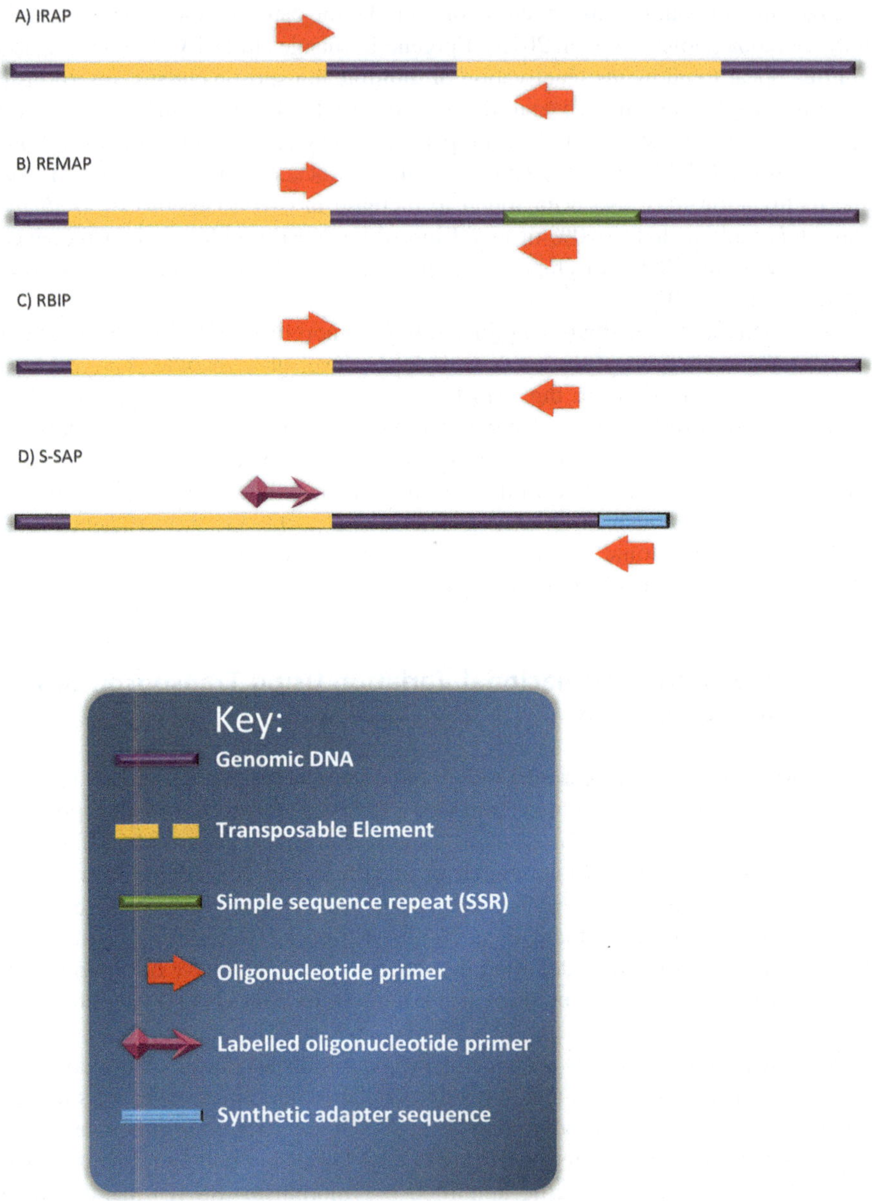

Fig. 1.1 Shows the different types of markers involving transposons discussed above

retrotransposons insertion sites (Kalendar et al. 1999; Kalendar and Schulman 2006). IRAP markers have been used to ensure genetic fidelity in banana, barley, wheat, and *Diospyros* (Muhammad and Othman 2005; Guo et al. 2006; Li et al. 2007; Carvalho et al. 2010; Campbell et al. 2011). Retrotransposon-based insertion polymorphisms (REMAP) use primers specifically targeting transposon insertion sites and aim at testing presence or absence of insertions hence amplifying regions between microsatellites and adjacent retrotransposons (Carvalho et al. 2010; Lizamore 2013). REMAP aims at amplifying regions between microsatellites and adjacent retrotransposons (Lizamore 2013). All markers are represented in Fig. 1.1.

1.5.12 Next-Generation Sequencing (NGS) for the Determination of Somaclonal Variation

Regeneration of plants by tissue culture is stressful and can lead to both genetic and epigenetic changes which lead to phenotypic polymorphism. NGS-based analyses have been carried out in *Arabidopsis* (Jiang et al. 2011) and rice (Sabot et al. 2011; Miyao et al. 2012) to confirm the extent of genomic change which occurs during tissue culture for induction of somaclonal variation. An elevated genome-wide DNA sequence mutation rate was observed in *Arabidopsis* regenerated plantlets with base substitution as major genomic change but with no detectable change in transposable element reactivation (Jiang et al. 2011). In rice, single nucleotide polymorphism (SNPs), nucleotide changes (including insertions and deletions) as well as transpositions were observed (Sabot et al. 2011; Miyao et al. 2012; Zhang et al. 2014). In wheat, however, transposable elements were found to be the most likely factors causing somaclonal difference at the genome level (Bara Anek et al. 2016).

1.5.13 MicroRNA Involvement in Tissue Culture

MicroRNA or miRNA, first discovered in *Caenorhabditis elegans*, has strong gene regulatory activity. These miRNAs are noncoding, approximately 19–24 nucleotides in length which have been recognized to play crucial regulatory roles in many biological processes and pathways (Carrington and Ambros 2003). MiRNAs have been reported to be highly involved in plant defense especially in stress responses, plant development, hormonal signaling, and seed germination. Recently miRNAs have been reported to be involved during in vitro culture of plants (Miguel and Marum 2011; Rodriguez-Eneiquez et al. 2011). Somaclonal variation is thought to be caused by both genetic and epigenetic factors which relate to transposable element transposition and small RNA-directed methylation (Smulders and De Klerk 2011; Neelakandan and Wang 2012).

Differential microRNA production (namely Mir169a and miR390) has been recorded in strawberry plantlets regenerated by tissue culture and these have been correlated with existing differences with the phenotype (Li et al. 2012). Regulatory roles of miRNAs have been highlighted during somatic embryogenesis of many plants (Nodine and Bartel 2010; Su et al. 2015; Zhang et al. 2015a, b) as well as their involvement in gene regulation of epigenetic processes. However, to date, no direct relationship has been found between somaclonal variation and microRNA. Further studies are required to characterize the miRNA expressed during somaclonal variation from different plants so that diagnostic miRNA associated with this phenomenon can be compiled and their possible use as biomarkers facilitated.

1.6 Conclusion

Regenerating plants by tissue culture may be aimed at maintaining the genetic integrity and the production of true to types but at times somaclonal variation may be desirable to increase the gene pool of a population with a narrow genetic base. Various methods for the detection of somaclonal variation have been reviewed and methods for assessing variant genomes at the molecular level allow for early and easy selection during a breeding experiment. These methods may be used in isolation but preferably in concertation to ascertain genetic fidelity.

References

Abdellatif MF, Hegazy AE, Aboshama HM, Emara HA, El-Shahed AA (2012) Morphological and molecular characterization of somaclonal variations in tissue culture-derived banana plants. J Genet Eng Biotechnol 10:47–53

Abreu IS, Carvalho CR, Clarindo WR (2014) Massal induction of *Carica papaya* L. 'golden' somatic embryos and somaclone screening by flow cytometry and cytogenetic analysis. Cytologia 79:475–484

Adkins SW, Kunanuvatchaidach R, Godwin ID (1995) Somaclonal variation in rice: drought tolerance and other agronomic characters. Aust J Bot 43:201–209

Afrasiab H, Iqbal J (2012) Biochemical and molecular characterization of somaclonal variants and induced mutants of potato (*Solanum tuberosum* L.) CV. Desiree. Pak J Bot 44:1503–1508

Agarwal M, Shrivastava N, Padh H (2008) Advances in molecular marker techniques and their applications in plant sciences. Plant Cell Rep 27:617–631

Alvarez ME, Nota F, Cambiagno DA (2010) Epigenetic control of plant immunity. Mol Plant Pathol 11:563–576

Amberger LA, Shoemaker RC, Palmer RG (1992) Inheritance of two independent isozyme variants in soybean plants derived from tissue culture. Theor Appl Genet 84(5–6):600–607

Arihara A, Kita T, Igarashi S, Goto M, Irikura Y (1995) White baron: a non-browning somaclonal variant of Danshakuimo (Irish cobbler). Am J Potato Res 72:701–705

Arun B, Joshi AK, Chand R, Singh BD (2003) Wheat somaclonal variants showing earliness, improved spot blotch resistance and higher yield. Euphytica 132:235–241

Asif MJ, Othman RY (2005) Characterization of fusarium wilt-resistant and fusarium wilt-susceptible somaclones of banana cultivar rastali (Musa AAB) by random amplified polymorphic DNA and retrotransposon markers. Plant Mol Biol Report 23:241–249

Baer G, Yemets A, Stadnichuk N, Rakhmetov D, Blume Y (2007) Somaclonal variability as a source for creation of new varieties of finger millet (*Eleusine coracana* (L.) Gaertn.). Cytol Genet 41:204–208

Bairu MW, Aremu AO, Van Staden J (2011) Somaclonal variation in plants: causes and detection methods. Plant Growth Regul 63:147–173

Bara Anek M, Cechova J, Kovacs T, Eichmeier A, Wang S, Raddova J et al (2016) Use of combined MSAP and NGS techniques to identify differentially methylated regions in Somaclones: a case study of two stable somatic wheat mutants. PLoS One 11(10):e0165749

Bartoszewski G, Havey MJ, Ziółkowska A, Długosz M, Malepszy S (2007) The selection of mosaic (MSC) phenotype after passage of cucumber (*Cucumis sativus* L.) through cell culture—a method to obtain plant mitochondrial mutants. J Appl Genet 1:1–9

Berenyi M, Gichuki ST, Schmidt J, Burg K (2002) Ty1-copia retrotransposon-based S-SAP (sequence-specific amplified polymorphism) for genetic analysis of sweet potato. Theor Appl Genet 105:862–869

Bhatia R, Singh KP, Jhang T, Sharma TR (2009) Assessment of clonal fidelity of micropropagated gerbera plants by ISSR markers. Sci Hortic 119:208–211

Biswas MK, Dutt M, Roy UK, Islam R, Hossain M (2009) Development and evaluation of in vitro somaclonal variation in strawberry for improved horticultural traits. Sci Hortic 122:409–416

Bobadilla Landey R, Cenci A, Georget F, Bertrand B et al (2013) High genetic and epigenetic stability in *Coffea arabica* plants derived from embryogenic suspensions and secondary embryogenesis as revealed by AFLP, MSAP and the phenotypic variation rate. PLoS One 8 (2):e56372

Brar DS, Jain SM (1998) Somaclonal variation: mechanism and application in crop improvement. In: Jain SM, Brar DS, Ahloowalia BS (eds) Somaclonal variation and induced mutations in crop improvement. Kluwer Academic Publishers, Dordrecht, pp 15–37

Brown PTH, Göbel E, Lörz H (1991) RFLP analysis of *Zea-mays* callus cultures and their regenerated plants. Theor Appl Genet 81:227–232

Campbell BC, LeMare S, Piperidis G, Godwin ID (2011) IRAP, a retrotransposon- based marker system for the detection of somaclonal variation in barley. Mol Breed 27:193–206

Carini F, De Pasquale F (2003) Micropropagation of *Citrus*. In: Mohan S, Ishii K (eds) Micropropagation of wood tree and fruits, vol 75. Kluwer Academic Publishers, London, pp 589–619

Carrington JC, Ambros V (2003) Role of microRNAs in plant and animal development. Science 301:336–338

Carvalho A, Guedes Pinto H, Martins Lopes P, Lima Brito J (2010) Genetic variability of old Portuguese bread wheat cultivars assayed by IRAP and REMAP markers. Ann Appl Biol 156:337–345

Cassells AC, Curry RF (2001) Oxidative stress and physiological, epigenetic and genetic variability in plant tissue culture: implications for micropropagators and genetic engineers. Plant Cell Tissue Org Cult 64:145–157

Chawla HS (2000) Introduction to plant biotechnology. Science Publishers, Inc., Enfield

Chen WH, Chen TM, Fu YM, Hsieh RM, Chen WS (1998) Studies on somaclonal variation in Phalaenopsis. Plant Cell Rep 18:7–13

Cloutier S, Landry B (1994) Molecular markers applied to plant tissue culture. In Vitro Cell Dev Biol Plant 30:32–39

Coggins LW, O'Prey M (1989) DNA tertiary structures formed in vitro by misaligned hybridization of multiple tandem repeat sequences. Nucleic Acids Res 17:7417–7426

D'Amato F (1985) Cytogenetics of plant cells and tissue cultures and their regenerates. Plant Sci 3:73–112

Damasco OP, Godwin ID, Smith MK, Adkins SW (1996) Gibberellic acid detection of dwarf off-types in micropropagated Cavendish bananas. Aust J Exp Agric 36:237–341

Dey T, Saha S, Ghosh PD (2015) Somaclonal variation among somatic embryo derived plants-evaluation of agronomically important somaclones and detection of genetic changes by RAPD in *Cymbopogon winterianus*. S Afr J Bot 96:112–121

Dorffling K, Melz G (1997) Improvement of frost tolerance in winter wheat by in vitro selection of proline over producing mutants. Acta Agronom Hungar 45:295–299

Duncan RR (1996) Tissue culture induced variation and crop improvement. In: Donald LS (ed) Advances in agronomy, vol 58. Academic Press, Waltham, pp 201–240

Edallo S, Zucchinali C, Perenzin M, Salamini F (1981) Chromosomal variation and frequency of spontaneous mutation associated with in vitro culture and plant regeneration in maize. Maydica 26:39–56

Evans DA, Sharp WR (1983) Single gene mutations in tomato plants regenerated from tissue culture. Science 221:949–951

Francischini JHMB, Kemper EL, Costa JB, Manechini JRV, Pinto LR (2017) DNA methylation in sugarcane somaclonal variants assessed through methylation- sensitive amplified polymorphism. Genet Mol Res 16(2):gmr16029585

Gao X, Yang D, Cao D, Ao M, Sui X, Wang Q et al (2010) In vitro micropropagation of *Freesia hybrida* and the assessment of genetic and epigenetic stability in regenerated plantlets. J Plant Growth Regul 29:257–267

Giorgetti L, Castiglione M, Turrini A, Ronchi V, Geri C (2011) Cytogenetic and histological approach for early detection of "mantled" somaclonal variants of oil palm regenerated by somatic embryogenesis: first results on the characterization of regeneration system. Caryologia 64:223–234

Gonzalez AI, Saiz A, Acedo A, Ruiz ML et al (2013) Analysis of genomic DNA methylation patterns in regenerated and control plants of rye (*Secale cereale* L.). Plant Growth Regul 70:227–236

Graebe J (2003) Gibberellin biosynthesis and control. Annu Rev Plant Physiol Plant Mol Biol 38:419–465

Guo D, Zhang H, Luo Z (2006) Genetic relationships of *Diospyros kaki* Thunb. And related species revealed by IRAP and REMAP analysis. Plant Sci 170:528–533

Gupta PK (1998) Chromosomal basis of somaclonal variation in plants. In: Jain SM, Brar DS, Ahloowalia BS (eds) Somaclonal variation and induced mutations in crop improvement. Kluwer Academic Publishers, Dordrecht, pp 149–168

Halušková J, Čellárová E (1997) RFLP analysis of *Hypericum perforatum* L. somaclones and their progenies. Euphytica 2:229–235

Hang A, Bregitzer P (1993) Chromosomal variations in immature embryo-derived calli from six barley cultivars. J Hered 84:105–108

Hao YI, Deng XX (2002) Occurrence of chromosomal variations and plant regeneration from long-term-cultured *citrus* callus. In Vitro Cell Dev Biol Plant 38:472–476

Hautea DM, Molina GC, Balatero CH, Coronado NB, Perez EB, Alvarez MTH, Canama AO, Akuba RH, Quilloy RB, Frankie RB, Caspillo CS (2004) Analysis of induced mutants of Philippine bananas with molecular markers. In: Jain SM, Swennen R (eds) Banana improvement: cellular, molecular biology, and induced mutations. Science Publishers, Inc., Enfield, pp 45–58

He XJ, Chen T, Zhu JK (2011) Regulation and function of DNA methylation in plants and animals. Cell Res 21:442–465

Hirochika H (1993) Activation of tobacco retrotransposons during tissue culture. EMBO J 12:2521–2528

Hirochika H, Sugimoto K, Otsuki Y, Tsugawa H, Kanda M (1996) Retrotransposons of rice involved in mutations induced by tissue culture. Proc Natl Acad Sci USA 93:7783–7788

Holland AJ, Cleveland DW (2009) Boveri revisited: chromosomal instability, aneuploidy and tumorigenesis. Nat Rev Mol Cell Biol 10:478–487

Hollister JD, Gaut BS (2009) Epigenetic silencing of transposable elements: a trade-off between reduced transposition and deleterious effects on neighboring gene expression. Genome Res 19:1419–1428

Huang J, Zhang K, Shen Y, Huang Z, Li M, Tang D, Gu M, Cheng Z (2009a) Identification of a high frequency transposon induced by tissue culture, nDaiZ, a member of the hAT family in rice. Genomics 93:274–281

Huang WJ, Ning GG, Liu GF, Bao MZ (2009b) Determination of genetic stability of long-term micropropagated plantlets of *Platanus acerifolia* using ISSR markers. Biol Plant 53:159–163

Isah T (2015) Adjustments to in vitro culture conditions and associated anomalies in plants. Acta Biol Cracov Ser Bot 57:9–28

Islam MA, Kloppstech K, Jacobsen HJ (2004) Efficient procedure for in vitro microrhizome induction in *Curcuma longa* L. (Zingiberaceae)–a medicinal plant of tropical Asia. Plant Tiss Culture 14:123–134

Jaligot E, Beulé T, Rival, A (2002) Methylation-sensitive RFLPs: characterisation of two oil palm markers showing somaclonal variation-sensitive associated polymorphism. Theor Appl Genet 104:1263–1269

Jiang C, Mithani A, Gan X, Belfield EJ, Klingler JP et al (2011) Regenerant *Arabidopsis* lineages display a distinct genome-wide spectrum of mutations conferring variant phenotypes. Curr Biol 21:1385–1390

Jin S, Mushke R, Zhu H, Tu L, Lin Z, Zhang Y, Zhnag S (2008) Detection of somaclonal variation of cotton (*Gossypium hirsutum*) using cytogenetics, flow cytometry and molecular markers. Plant Cell Rep 27:1303–1316

Kaeppler SM, Phillips RL (1993) DNA methylation and tissue culture-induced DNA methylation variation in plants. In Vitro Cell Dev Biol Plant 29:125–130

Kaeppler SM, Phillips RL, Olhoft P (1998) Molecular basis of heritable tissue culture-induced variation in plants. In: Jain SM et al (eds) Somaclonal variation and induced mutations in crop improvement. Kluwer Academic Publishers, Dordrecht, Netherlands, pp 465–484

Kaeppler SM, Kaeppler HF, Rhee Y (2000) Epigenetic aspects of somaclonal variation in plants. Plant Mol Biol 43:179–188

Kalendar R, Schulman AH (2006) IRAP and REMAP for retrotransposon-based genotyping and fingerprinting. Nat Protoc 1:2478–2484

Kalendar R, Grob T, Regina M, Suoniemi A, Schulman AH (1999) IRAP and REMAP: two new retrotransposon-based DNA fingerprinting techniques. Theor Appl Genet 98:704–711

Karp A (1994) Origins, causes and uses of variation in plant tissue cultures. In: Vasil IK, Thorpe TA (eds) Plant cell and tissue culture. Kluwer Academic Publishers, Dordrecht, pp 139–152

Khai TH, Lang NT (2005) Using SSR marker to identify allele variation of Somaclonal mutants in indica rice. Omon Rice 13:121–125

Kitimu SR, Taylor J, March TJ, Tairo F, Wilkinson MJ, Rodríguez-López CM (2015) Meristem micropropagation of cassava (*Manihot esculenta*) evokes genome-wide changes in DNA methylation. Front Plant Sci 6:590

Kunert K, Baaziz M, Cullis C (2003) *Techniques* for determination of *true-to-type date palm (Phoenix dactylifera L.)* plants: a literature review. Emirat J Agricult Sci 15:1–16

Labra M, Imazio S, Grassi F, Rossoni M, Sala F (2004) Vine-1 retrotransposon-based sequence-specific amplified polymorphism for *Vitis vinifera* L. genotyping. Plant Breed 123:180–185

Larkin PJ (1998) In: Jain SM, Brar DS, Ahloowalia BS (eds) Somaclonal variation and induced mutations in crop improvement. Kluwer Academic Publishers, Dordrecht, pp 3–13

Larkin PJ, Scowcroft WR (1981) Somaclonal variation a new source of variability from cell cultures for crop improvement. Theor Appl Genet 60:197–214

Leva AR, Petruccelli R, Rinaldi LMR (2012) Somaclonal variation in tissue culture: a case study with olive. In: Recent advances in plant in vitro culture. IntechOpen, London, pp 123–150

Levall MW, Bengtsson K, Nilsson NO, Hjerdin A, Halldén C (1994) Molecular characterization of UV-treated sugar beet somaclones using RFLP markers. Physiol Plant 90:216–220

Levinson G, Gutman GA (1987) Slipped-strand mispairing: a major mechanism for DNA sequence evolution. Mol Biol Evol 4:203–222

Li X, Yu X, Wang N, Feng Q, Dong Z, Liu L, Shen J, Liu B (2007) Genetic and epigenetic instabilities induced by tissue culture in wild barley (*Hordeum brevisubulatum* (Trin.) link). Plant Cell Tissue Org Cult 90:153–168

Li H, Zhao X, Dai H et al (2012) Tissue culture responsive micro RNAs in strawberry. Plant Mol Biol Report 30:1047

Linacero R, Rueda J, Esquivel E, Bellido A, Domingo A, Vazquez A (2011) Genetic and epigenetic relationship in rye, *Secale cereal* L, somaclonal variation within somatic embryo-derived plants. In Vitro Cell Dev Biol Plant 47:618–628

Lippman ZB, Grendel AV, Black M et al (2004) Role of transposable elements in heterochromatin and epigenetic control. Nature 430:471–476

Lisch D (2009) Epigenetic regulation of transposable elements in plants. Annu Rev Plant Biol 60:43–66

Lizamore DK (2013) A Study of endogenous transposon activity in Grapevine (*Vitis vinifera* L.). PhD thesis. Lincoln University, Christchurch, New Zealand

Lorz H, Scowcroft W (1983) Variability among plants and their progeny regenerated from protoplasts of Su/su heterozygotes of *Nicotiana tabacum*. Theor Appl Genet 66:67–75

Makarevitch I, Waters AJ, West PT, Stitzer M, Hirsch CN, Ross-Ibarra J, Springer NM (2015) Transposable elements contribute to activation of maize genes in response to abiotic stress. PLoS Genet 11:e1004915

Marum L, Rocheta M, Maroco J, Oliveira MM, Miguel C (2009) Analysis of genetic stability at SSR loci during somatic embryogenesis in maritime pine (*Pinus pinaster*). Plant Cell Rep 28:673–682

Matthes M, Singh R, Karp A (2001) Variation in oil palm (*Elaeis guineensis* Jacq.) tissue culture-derived regenerants revealed by AFLPs with methylation-sensitive enzymes. Theor Appl Genet 102:971–979

Meudt HM, Clarke AC (2007) Almost forgotten or latest practice? AFLP applications, analyses and advances. Trends Plant Sci 12:106–117

Miguel C, Marum L (2011) An epigenetic view of plant cells cultured in vitro: somaclonal variation and beyond. J Exp Bot 62:3713–3725

Miyao A, Nakagome M, Ohnuma T, Yamagata H, Kanamori H et al (2012) Molecular spectrum of somaclonal variation in regenerated rice revealed by whole-genome sequencing. Plant Cell Physiol 53:256–264

Mohanty S, Panda MK, Sahoo S, Nayak S (2011) Micropropagation of *Zingiber rubens* and assessment of genetic stability through RAPD and ISSR. Biol Plant 55:16–20

Muhammad AJ, Othman RY (2005) Characterization of *Fusarium* wilt-resistant and *Fusarium* wilt-susceptible somaclones of banana cultivar rastali (*Musa* AAB) by random amplified polymorphic DNA and retrotransposon markers. Plant Mol Biol Report 23:241–249

Mujib A, Banerjee S, Dev Ghosh P (2007) Callus induction, somatic embryogenesis and chromosomal instability in tissue culture-raised hippeastrum (*Hippeastrum hybridum* cv. United Nations). Propagat Ornament Plant 7:169–174

Müller E, Brown PTH, Hartke S, Lörz H (1990) RFLP-analysis of rice plants regenerated from tissue cultures. In: Nijkamp HJJ et al (eds) Progress in plant cellular and molecular biology. Kluwer Academic Publication, Dordrecht, pp 153–156

Neelakandan A, Wang K (2012) Recent progress in the understanding of tissue culture-induced genome level changes in plants and potential applications. Plant Cell Rep 31:597–620

Ngezahayo F, Dong Y, Liu B (2007) Somaclonal variation at the nucleotide sequence level in rice (*Oryza sativa* L.) as revealed by RAPD and ISSR markers and by pairwise sequence analysis. J Appl Genet 48:329–336

Nodine MD, Bartel DP (2010) MicroRNAs prevent precocious gene expression and enable pattern formation during plant embryogenesis. Genes Dev 24:2678–2692

Nookaraju A, Agrawal D (2012) Genetic homogeneity of in vitro raised plants of grapevine cv. Crimson Seedless revealed by ISSR and microsatellite markers South African. J Bot 78:302–306

Orzechowska M, Stepien K, Kaminska T, Siwinska D (2013) Chromosome variations in regenerants of *Arabidopsis thaliana* derived from 2-and 6-week-old callus detected using flow cytometry and FISH analysis. Plant Cell Tissue Org Cult 112:263–273

Osipova ES, Kokaeva ZG, Troitskij AV, Dolgikh YI, Shamina ZB, Gotimskij SA (2001) RAPD analysis of maize somaclones. Genetika-Moskva 37:91–96

Peschke VM, Phillips RL (1991) Activation of the maize transposable element *Suppressor-mutator* (*Spm*) in tissue culture. Theor Appl Genet 81:90–97

Phinney BO (1985) Gibberellin A1 dwarfism and shoot elongation in higher plants. Biol Plant 27:172–179

Pinto G, Loureiro J, Lopes T, Santos C (2004) Analysis of the genetic stability of *Eucalyptus globulus Labill* somatic embryos by flow cytometry. Theor Appl Genet 109:580–587

Polanco C, Ruiz ML (2002) AFLP analysis of somaclonal variation in *Arabidopsis thaliana* regenerated plants. Plant Sci 162:817–824

Pontaroli AC, Camadro EL (2005) Somaclonal variation in *Asparagus officinalis* plants regenerated by organogenesis from long term callus cultures. Genet Mol Biol 28:423–430

Rajora OP, Rahman MH (2003) Microsatellite DNA and RAPD fingerprinting, identification and genetic relationships of hybrid poplar (*Populus x canadensis*) cultivars. Theor Appl Genet 106:470–477

Rakocevic A, Mondy S, Tirichine L, Cosson V, Brocard L, Iantcheva A et al (2009) MERE1, a low-copy-number copia-type retroelement in *Medicago truncatula* active during tissue culture. Plant Physiol 151:1250–1263

Rastogi J, Siddhant PB, Sharma BL (2015) Somaclonal variation: a new dimension for sugarcane improvement. GERF Bullet Biosci 6:5–10

Ray T, Dutta I, Saha P, Das S, Roy SC (2006) Genetic stability of three economically important micropropagated banana (*Musa* spp.) cultivars of lower Indo-Gangetic plains, as assessed by RAPD and ISSR markers. Plant Cell Tissue Org Cult 85:211–214

Rodrigues PHV, Tulmann Neto A, Cassieri Neto P, Mendes BMJ (1998) Influence of the number of subcultures on somoclonal variation in micropropagated Nanico (Musa spp., AAA group). Acta Hortic 490:469–473

Rodriguez-Eneiquez J, Dickinson HG, Grant-Downton RT (2011) MicroRNA misregulation: an overlooked factor generating somaclonal variation. Trends Plant Sci 16:242–248

Rosales-Serna R, Hernandez-Delgado S, Gonzalez-Paz M, Acosta-Gallegos A, Mayek-Perez N (2005) Genetic relationships and diversity revealed by AFLP markers in Mexican common bean bred cultivars. Crop Sci 45:1951–1957

Sabot F, Picault N, El-Baidouri M, Llauro C, Chaparro C et al (2011) Transpositional landscape of the rice genome revealed by paired-end mapping of high-throughput re-sequencing data. Plant J 66:241–246

Saker M, Bekheet S, Taha HS, Moursy HA (2000) Detection of somaclonal variation in tissue culture-derived date palm plants using isozyme analysis and RAPD fingerprints. Biol Plant 43:347–351

Salman RM, Al Jibouri AAM, Al Quadhy WK, Omar MS (1988) Isozyme and chromosomal analyses of tissue culture derived date palms. Date Palm J 6:401–411

Sanchez-Teyer LF, Quiroz-Figueroa F, Loyola-Vargas V, Infante D (2003) Culture-induced variation in plants of *Coffea arabica* cv. Caturra rojo, regenerated by direct and indirect somatic embryogenesis. Mol Biotechnol 23:107–115

Sandoval J, Kerbellec F, Côte F, Doumas P (1995) Distribution of endogenous gibberellins in dwarf and giant off-types banana (*Musa* AAA, cv. Grand Nain) plants from in vitro propagation. Plant Growth Regul 17:219–224

Sato M, Kawabe T, Hosokawa M, Tatsuzawam F, Doi M (2011) Tissue culture induced flower-color changes in Saintpaulia caused by excision of the transposon inserted in the flavonoid 39, 59 hydroxylase (F3959H) promoter. Plant Cell Rep 30:929–939

Schellenbaum P, Mohler V, Wenzel G, Walter B (2008) Variation in DNA methylation patterns of grapevine somaclones (*Vitis vinifera* L.). BMC Plant Biol 8:78

Slazak B, Sliwinska E, Saługa M, Ronikier M, Bujak J, Słomka A, Goransson E, Kuta E (2015) Micropropagation of *Viola uliginosa* (Violaceae) for endangered species conservation and for somaclonal variation-enhanced cyclotide biosynthesis. Plant Cell Tissue Org Cult 120:179–190

Smulders M, de Klerk G (2011) Epigenetics in plant tissue culture. Plant Growth Regul 63:137–146

Smýkal P, Valledor L, Rodriguez R, Griga M (2007) Assessment of genetic and epigenetic stability in long-term in vitro shoot culture of pea (*Pisum sativum* L.). Plant Cell Rep 26:1985–1998

Soniya EV, Banerjee NS, Das MR (2001) Genetic analysis of somaclonal variation among callus-derived plants of tomato. Curr Sci 80:1213–1215
Su YH, Liu YB, Zhou C, Li XM, Zhang XS (2015) The microRNA167 controls somatic embryogenesis in *Arabidopsis* through regulating its target genes ARF6 and ARF8. Plant Cell Tissue Org Cult 124:405–417
Tam SM, Mhiri C, Vogelaar A, Kerkveld M, Pearce SR, Grandbastien MA (2005) Comparative analyses of genetic diversities within tomato and pepper collections detected by retrotransposon-based SSAP, AFLP and SSR. Theor Appl Genet 110:819–831
Tang CY (2005) Somaclonal variation a tool for improvement of Cavendish banana cultivars. Acta Hortic 692:61–65
Tawar PN, Sawant RA, Dalvi SG, Nikam AA, Tawar PG, Devarumath RM (2008) An assessment of somaclonal variation in micropropagated plants of sugarcane by RAPD markers. Sugar Technol 10:124–127
Taylor PWJ, Geijskes JR, Ko HL, Fraser TA, Henry RJ, Birch RG (1995) Sensitivity of random amplified polymorphic DNA analysis to detect genetic change in sugarcane during tissue culture. Theor Appl Genet 90:1169–1173
Thieme R, Griess H (2005) Somaclonal variation in tuber traits of potato. Potato Res 48:153–165
Thomas J, Vijayan D, Joshi SD, Lopez SJ, Kumar RR (2006) Genetic integrity of somaclonal variants in tea [*Camellia sinensis* (L.) O Kuntze] as revealed by inter simple sequence repeats. J Biotechnol 132:149–154
Trujillo I, Garcia E (1996) Strategies for obtaining somaclonal variants resistant to yellow Sigatoka (*Mycosphaerella musicola*). Infomusa 5:212–213
Us-Camas R, Rivera-Solis G, Duarte-Ake F, Dela Pena C (2014) In vitro culture: An epigenetic challenge for plants. Plant Cell Tissue Org Cult 118:187–201
Vijayan K, Chatterjee SN (2003) ISSR profiling of Indian cultivars of mulberry (*Morus* spp.) and its relevance to breeding programs. Euphytica 131:53–63
Vos P, Hogers R, Bleeker M, Reijans M, Lee TVD, Hornes M et al (1995) AFLP: a new technique for DNA fingerprinting. Nucleic Acids Res 23:4407–4414
Wang Y, Wang F, Zhai H, Liu Q (2007) Production of a useful mutant by chronic irradiation in sweet-potato. Sci Hortic 111:173–178
Waugh R, McLean K, Flavell A, Pearce S, Kumar A, Thomas B, Powell W (1997) Genetic distribution of bare 1-like retrotransposable elements in the barley genome revealed bysequence-specific amplification polymorphisms (S-SAP). Mol Gen Genet 253:687–694
Wondyifraw T, Wannakrairoj S (2004) Micropropagation of Krawan (*Amomum krervanh* Pierre ex Gagnep). Sci Asia 30:9–15
Yaish MW, Peng M, Rothstein SJ (2014) Global DNA methylation analysis using methyl-sensitive amplification polymorphism (MSAP). Methods Mol Biol 1062:285–298
Yang H, Tabei Y, Kamada H, Kayano T, Takaiwa F (1999) Detection of somaclonal variation in cultured rice cells using digoxigenin-based random amplified polymorphic DNA. Plant Cell Rep 18:520–526
Zaid A, Al Kaabi H (2003) Plant-off types in tissue culture-derived date palm (*Phoenix dactylifera* L.). Emirat J Agricult Sci 15:17–35
Zhang M, Wang H, Dong Z, Qi B, Xu K, Liu B (2010) Tissue culture induced variation at simple sequence repeats in sorghum (*Sorghum bicolor* L.) is genotype-dependent and associated with down-regulated expression of a mismatch repair gene, MLH3. Plant Cell Rep 29:51–59
Zhang D, Wang Z, Wang N, Gao Y, Liu Y, Wu Y et al (2014) Tissue culture-induced heritable genomic variation in rice, and their phenotypic implications. PLoS One 9:e96879
Zhang C, Cao L, Rong L, An Z, Zhou W, Ma J, Shen WH, Zhu Y, Dong A (2015a) The chromatin-remodeling factor AtINO80 plays crucial roles in genome stability maintenance and in plant development. Plant J 82:655–668
Zhang F, Dong W, Huang L, Song A, Wang H, Fang W, Chen F, Teng N (2015b) Identification of microRNAs and their targets associated with embryo abortion during *Chrysanthemum* cross breeding via high-throughput sequencing. PLoS One 10:e0124371

Synthetic Seed Technology in Some Ornamental and Medicinal Plants: An Overview

2

Mehpara Maqsood, Mir Khusrau, A. Mujib, and Zahoor A. Kaloo

Abstract

Synthetic seed technology has been successfully employed in various medicinal and ornamental plants. It has immense potential in preserving callus, somatic embryo (SE), and other important tissues of valuable germplasm. The method can also be exploited as a tool of multiplication in diverse plant groups. Synthetic seed production in plants facilitates season-independent propagation of elite germplasm, accelerates seedless, ornamental, and ploidy plants multiplication with unique traits. Synthetic seed/encapsulation approach has a high beneficiary role in improved and fast transportation of propagules. The obligatory requirement for application of artificial seed technology en masse involves high-scale production of quality micro-propagules, which nowadays is a serious limiting factor. There are numerous factors controlling the success of development and germination of synthetic seeds. These include matured quality embryos, temperature, storage time, sodium alginate percentage, calcium chloride concentration, hardening time, and plant growth regulators (PGRs) concentrations. Encapsulation technology is an emerging area to open new vista in conserving, propagating, and shipment of valuable plant material for research affecting floriculture and pharmaceutical industry. In this chapter, the artificial seed technique status is reviewed in some plants of ornamental and medicinal importance.

M. Maqsood
Govt. College for Women, Srinagar, Jammu & Kashmir, India

M. Khusrau
Govt. Degree College (Boys), Anantnag, Jammu & Kashmir, India

A. Mujib (✉)
Department of Botany, Jamia Hamdard, New Delhi, India

Z. A. Kaloo
University of Kashmir, Hazratbal, Srinagar, Jammu & Kashmir, India

I. Siddique (ed.), *Propagation and Genetic Manipulation of Plants*,
https://doi.org/10.1007/978-981-15-7736-9_2

Keywords

Encapsulation · Germplasm · Matrix · Medicinal plants · Ornamental

2.1 Introduction

The production of seed is an outcome of usual sexual process; in cross-pollinating plants, however, the produced seeds are genetically different from parental stock (Senaratna 1992). An important seed characteristic is its ability to undergo dormancy by self-drying, enabling seeds to be preserved for a long time. This is primarily due to plants' vegetative nature in which the distribution and exchange of seeds become inconvenient and a threat of disease dissemination is quite high (Chaudhury and Malik 2003). Recently, an increasing interest is noted on preparation of artificial seed through encapsulation of seeds/other plant parts—an important way of preservation and exchange of germplasm (Gantait et al. 2015). This synthetic seed technology is of immense importance in mass propagation of elite plant species where sexual seed propagation is plagued with incompatibilities. Frequently, the somatic embryos (SE) are used in encapsulation study; in recent times, enough efforts have been given to encapsulate non-embryogenic and other vegetative parts like axillary buds, shoot tips, and nodal segments (Sundararaj et al. 2010; Verma et al. 2010). Synthetic/artificial seeds are analogous to botanical seeds in which SEs are covered with one or more artificial layers, looking like a capsule (Cangahuala-Inocente et al. 2007). These encapsulated embryos can be used in conservation purposes of important germplasm including endangered plants; it also helps facilitating the exchange of materials across the world research centers (Rai et al. 2009).

The concept of artificial seed is floated since the discovery of SE induction in various investigated plant species. Artificial seed, also known as synseed, is first described by Murashige (1977) as an encapsulated single somatic embryo. It was later defined as a system engineered for practical use in commercial plant production (Gray et al. 1991). Unfortunately, very little attention has been paid to encapsulate vegetative plant parts like calli, nodal segments, buds, and shoot apices (Zych et al. 2005). The successful production of artificial seeds containing embryos or other nonsexual plant parts can open up new vistas in medicinal and floriculture sectors. But the application of artificial seeds depends primarily on the quality of somatic embryos similar to zygotic embryos analogous in morphology, physiology, and biochemistry (Redenbaugh et al. 1988).

There are two types of synthetic seeds, i.e., desiccated and hydrated (Bhojwani and Razdan 2006). The desiccated types were first introduced as naked or encapsulated SEs in Polyox, followed by desiccation (Kitto and Janick 1982). Desiccation is applied slowly over 1 or 2 weeks using decreasing relative humidity or is petri dishes placed overnight on laminar airflow chamber bench rapidly (Ara et al. 2000).

The hydrated artificial seed was first developed by encasing embryos of *Medicago sativa* and the technique is used in plant species where SEs are recalcitrant and sensitive to temperature (Redenbaugh et al. 1984). These seeds are prepared by encapsulating SEs or other vegetative parts in a hydrogel. Various methods have been attempted in producing artificial seeds in which the use of calcium alginate is found most successful (Redenbaugh 1983). Both the hydrated and desiccated synthetic seeds are plagued with technical limitations. The gaseous exchange/embryo respiration is often an issue with hydrated synthetic, which also dried out rapidly unless provided with humid conditions. At the same time, desiccation methodology damages the rudimentary embryo within the capsule very fast reducing the application of synthetic seed concept (Redenbaugh 1983).

Since the first report of Steward et al. (1958) and Reinert (1958), the incidence of somatic embryogenesis is observed in a number of tropical plants. These include cereals, beverages, spices, fruits, legumes, fibers, etc. In most of the situations, only a few plants were produced and conversion frequencies, i.e., percentage of embryo forming plants was not reported at all (Redenbaugh 1983). For commercial viewpoint and for artificial seed technology to be successful, large-scale, high-quality embryo production and faster germination comparable to true seeds are absolutely essential. In ornamentals, in particular, tissue culture technique is applied successfully for plant propagation but in vitro embryogenesis research has not been taken off for production system. In other groups, however, this technology has been attempted to a wide number of plants of different families. The propagation and simultaneous conservation of pharmaceutically important herbs using axillary buds was earlier reported (Mandal et al. 2000). A method of producing encapsulable units of synthetic seeds was reported in *Asparagus officinalis* (Kanji and Yuji 2001). The vasaka (*Adhatoda vasica*) shoot buds were similarly encapsulated in Na alginate solution in forming artificial seeds (Anand and Bansal 2002).

The current effort aimed to give a brief description of artificial seed production in some select medicinal and ornamental plants, in which the technology is developed. The protocol of encapsulation is presented in the below scheme:

2.2 Preparation of Synthetic Seeds

Selection of explants for encapsulation.

↓

Preparation of MS medium.

↓

Surface sterilization of explants for encapsulation.

↓

Preparation of alginate (%) and calcium chloride (mM) solution.

↓

Encapsulation of explants in sodium alginate solution using a wide bored pipette tip.

↓

Dropped sodium alginate along with explants into calcium chloride.

↓

Kept on rotary shaker for some time.

↓

Washed with autoclaved double distilled water.

↓

Dried synthetic seeds with autoclaved paper.

↓

Placed seeds on suitable MS medium.

↓

Hardened the plant in greenhouse.

The sodium alginate and calcium chloride level varies with plant species. The duration of time to harden the seeds also varies with the different plant species. The mechanism involves the replacement of Na^+ of sodium alginate (water soluble) with Ca^+ of calcium chloride to form calcium chloride (water insoluble). In an important medicinal plant, *Catharanthus roseus*, sucrose was added to the alginate which functions as endosperm provided nourishment to the growing embryo (Mehpara et al. 2012).

2.3 Synthetic Seeds of Medicinal Plants

Several medicinal plants occurring naturally come under rare, endangered, and endemic category. This decline in number is due to habitat destruction by a variety of anthropogenic activities like overgrazing, unregulated harvesting, global warming, and pollution and also by diverse natural phenomena (fewer seed setting, dormancy, poor rate of germination owing to incompatibility, etc.). It is therefore important to propagate these plants fast and conserve medicinal plants in vitro along with other preservation methods. In medicinal plants, the duration of incubation temperature had direct influence on artificial seed conversion, i.e., longer storage decreased conversion frequency irrespective of temperature in *Tylophora indica* (Gantait et al. 2017). Similar promising outcome was reported in oil palm in which cold storage influences metabolic rate in artificial seeds, embryos remain in a quiescent state and aiding preservation (Palanyandy et al. 2015). *Rumex vesicarius* L. belonging to the family Polygonaceae is a wild edible medicinal herb, propagated through seeds. The seeds are called dormant/recalcitrant with low seed germination rate (20%) which lessens its survivability (Asrar 2011). In such situation, the induction of mass-scale embryo formation is essential and encapsulation of SE by forming synthetic seed has huge applications. This methodology is noted to be highly successful in many plants like *Adhatoda vesica* (Anand and Bansal 2002), *Allium sativa* (Bekheet 2006), and *Daucus carota* (Latif et al. 2007). In *C. roseus*, an encapsulation process was developed for making synthetic seeds by using sodium

Table 2.1 Some important medicinal plants and various explants used for making synthetic seeds (Courtesy and source: Reddy et al. 2012; Gantait et al. 2015)

Medicinal plants	Explants used	Citations
Catharanthus roseus *Adhatoda vasica*	*SE* *Shoot buds*	Mehpara et al. (2012) Anand and Bansal (2002)
Allium satium L.	*Bulblets*	Bekheet (2006)
Bacopa monnieri L.	*Nodal microcutting*	Ramesh et al. (2009)
Cannabis sativa L.	*Nodal segments*	Lata et al. (2011)
Coriandrum sativum L.	*SE*	Stephen and Jayabalan (2000)
Cucumis sativus L.	*SE*	Tabassum et al. (2010)
Daucus carota L.	*SE*	Latif et al. (2007)
Helianthus annuus L.	*Shoot tip*	Katouzi et al. (2011)
Hyoscyamus muticus L.	*SE*	Pandey and Chand (2005)
Morus indica L.	*Axillary bud*	Kavyashree et al. (2006)
Rauvolfia serpentina (L.) Benth. Ex Kurz.	*Micro shoot*	Faisal et al. (2012)
Ocimum sanctum L.	*Axillary bud*	Mandal et al. (2000)
Simmondsia chinensis L.	*Shoot apex and axillary bud*	Hassan (2003)
Solanum nigrum L.	*Shoot tip*	Verma et al. (2010)
Valeriana wallichii	*Apical and axial shoot bud*	Mathur et al. (1989)
Ocimum basilicum	Axillary bud	Mandal et al. (2000)
Paulownia elongata	*SE*	Ipekci and Gozukirmizi (2003)
Chlorophytum borivilianum	*Shoot bud*	Dave et al. (2004)
Rhodiola kirilowii	Axillary bud and callus	Zych et al. (2005)
Withania somnifera	*Shoot tip*	Singh et al. (2006)

alginate (2.5%) + calcium chloride (100 mM) added solution. The encapsulated embryos germinated well and produced seedlings in NAA- and BAP-supplemented MS medium (Mehpara et al. 2012). The artificial synthetic seeds were preserved at low temperature (4 °C) for about 10 weeks, which later produced plantlets without losing germination ability.

Beside embryos, the use of other explants is reported for the production of artificial seeds in several studied medical plants like *Fragaria ananasa* and *Rubus idaeus* (Lisek and Orlikowska 2004), *Valeriana wallichii* (Mathur et al. 1989), *Rauvolfia serpentina* (Ray and Bhattacharya 2008), *Cineraria maritime* (Srivastava et al. 2009), *Cannabis sativa* (Lata et al. 2009), etc. The development of synthetic seeds in various medicinal plants is presented in Table 2.1:

In several studied plants like *R. vesicarius* L., the torpedo-shaped SEs were chosen for encapsulation (Nandini et al. 2014). The 2.0% alginate concentration in 100 mM $CaCl_2$ was optimum for encapsulation when kept for 45 min for complexation.

2.4 Synthetic Seeds and Ornamental Plants

A number of ornamentals/plants have desiccation-sensitive seeds showing storage time very limited (a few weeks to months). High pathogenic incidences make things more difficult. Synthetic seed procedure is an alternative way of in vitro propagation and preservation of elite germplasm including ornamentals. Ornamental plants constitute an important component in floriculture industry (Ali et al. 2007) and these groups composed of bulbous; tuberous; and foliage used in pot, landscaping, and house/office decorative purposes. Several protocols have been developed for the encapsulation of in vitro-regenerated SEs, protocorms, and other important explants (Lambardi et al. 2006). Some important ornamentals and the explants used for encapsulation are presented in Table 2.2.

The synthetic seed technology and conservation of important germplasm like medicinal, spices, and ornamental have been attempted in author's laboratory (Mujib et al. 2016; Ali et al. 2018). *Caladium,* an important monocotyledonous genus, belongs to the family Araceae. There are 12 species with about 2000 different cultivars. The foliage of *Caladium* is colored and very attractive; the leaves have green; white; red pigments, some are spotted, with red/green veinations, and are used as indoor decorative pot plants (Deng et al. 2007). In *Caladium,* the induction of SEs was indirect, i.e., callus was induced first from cultivated explants on which variable numbers of embryos were produced (Mujib et al. 2016). Compact callus cultured on

Table 2.2 Some ornamental plants and various explants used for the development of synthetic seeds (Courtesy and source: Reddy et al. 2012; Gantait et al. 2015)

Ornamental plants	Explants used	Citations
Caladium bicolor	SE	Mehpara et al. (2015)
Hyoscyamus muticus L.	SE	Pandey and Chand (2005)
Eustoma grandiflorum	SE	Ruffoni et al. (1994)
Genista monosperma	SE	Ruffoni et al. (1994)
Betula pendula	Soot tips	Piccioni and Standardi (1995)
Lilium longiflorum.	Bulbs	Standardi et al. (1995)
Syringa vulgaris	Axillary buds	Refouvelet et al. (1998)
Anthurium andreanum	Embryogenic calli	Nhut et al. (2004)
Gypsophila paniculata	Shoot tips	Rady and Hanafy (2004)
Saintpaulia ionantha	*Micro shoots*	Daud et al. (2008)
Nerium oleander	Axillary nodes	Ozden et al. (2008)
Photinia fraseri	Axillary nodes	Ozden et al. (2008)
Syringa vulgaris	Axillary nodes	Ozden et al. (2008)
Spathoglottis plicata	Protocorms	Khor et al. (1998)
Geodorum densiflorum	"Protocorm-like" body	Datta et al. (1999)
Cremastra appendiculata	"Protocorm-like" body	Zhang et al. (2009)
Coelogyne breviscapa	"Protocorm-like" body	Mohanraj et al. (2009)
Flickingeria nodosa.	"Protocorm-like" body	Nagananda et al. (2011)
Vanda coerulea	"Protocorm-like" body	Sarmah et al. (2010)

BAP- and NAA-added medium started to differentiate embryos in numbers. The embryo-forming ability (i.e., the embryo numbers/callus mass and embryogenesis frequency) however, varies with plant growth regulators used. On an average, three to five SEs were differentiated per callus mass. The embryo number gradually improved with time and in NAA + BAP-amended medium, good numbers of embryos were observed. The embryos progressed into matured embryo (swollen base and pointed tip) and these somatic embryos were selected for artificial seed production study. Mehpara et al. (2015) developed synthetic seeds by mixing various sodium alginate + calcium chloride levels in *Caladium bicolor*. The synthetic seeds/beads were developed by using embryos in sucrose (3.0%), sodium alginate (3.0%), and calcium chloride (100 mM) amended gel and successful plantlet recovery was obtained (Fig. 2.1d–f). The prepared artificial beads were kept at low temperatures at 4 °C, which showed good germination rate following 12 weeks of low-temperature storage. The conversion percentage (synthetic seed to plantlet regeneration) was, however, reduced with passage time.

2.5 Factors Affecting Development of Synthetic Seeds

The various factors that have noticeable effect on the formation of encapsulated seeds are explants used, percentage of sodium alginate, concentration of calcium chloride, duration for hardening, regulation of rotary rpm for hardening (Fig. 2.1a, b), plant growth regulators in culture medium, incubation time, etc. Various types of explants are used in the formation of synthetic seeds that govern their germination/viability:

2.6 Factors Affecting Synthetic Seeds' Performance

The viability of artificial seeds is influenced by several factors viz. storage time, temperature level, medium selection, and aseptic conditions (Fig. 2.1c). The germinability of stored synthetic seeds increased with high temperature and declined in cold storage in *R. vesicarius* L., whereas no significant difference was noted in room storage (Nandini et al. 2014). Another decisive factor is the quality of storage during transport, which regulates physiology of synthetic seeds and in turn determines the success of plantlet conversion from artificial seeds. Higher conversion efficiency of synthetic seeds was reported at cold (4 °C) temperature as was noted in *Cleopatra tangerine* (Nieves et al. 1998) and Apple stock (Sicurani et al. 2001). A promising degree of interaction was observed between cold storage (4 °C) and storage interval up to 45 days for regrowth as well as root and shoot development as reported in olive (Ikhlaq et al. 2010). The storage of encapsulated seed is a crucial factor for transport and conservation of germplasm on which the success of synthetic seeds depends.

Fig. 2.1 (**a**) Somatic embryos in *Catharanthus roseus,* (**b**) encapsulation of embryos, (**c**) emergence of plantlets from matrix in *C. roseus,* (**d**) somatic embryos in *Caladium bicolor,* (**e**) encapsulation of embryos, and (**f**) emergence of plantlets from matrix in *C. bicolor*

2.7 Advantages of Synthetic Seeds

- The synthetic seeds may be used as an alternative planting stock especially to forest and other economically important plant materials (Asmah et al. 2011).
- The use of artificial seeds offers additional avenues for direct planting to greenhouse or field, thus several intermediate steps may be bypassed (Reddy et al. 2012).
- Artificial seeds have merits in plants (cross-pollinated) where sexual incompatibility-based issues, like F1 seed setting and germination, are a major challenge (Latif et al. 2007).
- The synthetic seeds have multiple advantages including uniformity in production, higher scale-up capacity, potential for automation production process. Handling is simple and even during transportation; long-term preservation of valuable germplasm is also a reality (Singh et al. 2007).
- Synthetic seeds are exploited nowadays in advanced cryopreservation procedures with an aim at long-term preservation of germplasm (Ruby et al. 2015).

2.8 Disadvantages of Synthetic Seeds

Despite numerous advantages, synthetic seed possesses various limitations:

- Lack of efficient root systems in plantlets derived from synthetic seeds; limitations in artificial seed production and sowing of commercially important plants/varieties in aseptic cultural conditions (Hung and Trueman 2012).
- Scale-up production of micropropagules/embryos for making capsules in gel matrix at feasible rate is a serious limitation, and micropropagation system is still another major bottleneck in many woody plant genera, required for the development of artificial seed technology (Rihan et al. 2012).
- The planting of artificial seeds directly in soil or in other substances like soilrite, organic matter, vermiculite under aseptic conditions is a practical challenge to growers of developing countries (Jung et al. 2004).

2.9 Conclusion

In summary, the development of synthetic seed protocol and optimization of plant regeneration system will help in alternative propagation method of important plant genera like *Caladium* where plants do not propagate through seeds. Synchronous, uniform, and fast SE formation and encapsulation of embryos could provide a realistic approach for germplasm preservation. The encapsulated SEs were successfully preserved at low temperature (4 °C) for 10–12 weeks; further refinement of process may improve in enabling "short and medium" term storage possibility of in vitro-cultivated tissues of other horticulture and medicinal plants. Similarly, medicinal plants are important sources of alkaloids, conservation/cryopreservation of cell lines/embryogenic tissues with higher yield even up to medium term would extend its uses further in alkaloid enrichment study.

Acknowledgments The authors are indebted to the Department of Botany, Jamia Hamdard, New Delhi for receiving various supports.

References

Ali A, Munawar A, Naz S (2007) An in vitro study on micropropagation of *Caladium bicolor*. Int J Agric Biol 5:731–735

Ali M, Mujib A, Zafar N, Tonk D (2018) Somatic embryogenesis, biochemical alterations and synthetic seed development in two varieties of coriander (*Coriandrum sativum* L.). Adv Hortic Sci 32:239–248

Anand Y, Bansal YB (2002) Synthetic seed: a novel approach of in vitro plantlet formation in Vasaka (*Adhatoda vasica* Nees). Plant Biotechnol 19:159–162

Ara H, Jaiswal U, Jaiswal VS (2000) Synthetic seed: prospects and limitations. Curr Sci 78:1438–1444

Asmah HN, Hasnida HN, Zaimah NAN, Noraliza A, Salmi NN (2011) Synthetic seed technology for encapsulation and regrowth of in vitro derived *Acacia* hybrid shoot and axillary buds. Afr J Biotechnol 10:7820–7824

Asrar AA (2011) Seed germination induction of Hommaidh (*Rumex vesicarius* L.) by gibberlic acid and temperature applications. Eurasian J Agricult Environ Sci 10:310–317

Bekheet SA (2006) A synthetic seed method through encapsulation of in vitro proliferated bulblets of garlic (*Allium sativum* L). Arab J Biotechnol 9:415–426

Bhojwani SS, Razdan MK (2006) Plant tissue culture: theory and practice. Elsevier Science, Netherland, pp 125–166

Cangahuala-Inocente GC, Dal Vesco LL, Steinmacher D, Torrees AC, Guerra MP (2007) Improvements in somatic embryogenesis protocol in Feijoa (*Acca sellowiana* (berg) biuret). Induction, conversion and synthetic seeds. Sci Hortic 111:228–234

Chaudhury R, Malik SK (2003) Strategies for achieving short, medium- and long-term conservation of desiccation-sensitive seeds. In: Chaudhury R, Pandey R, Malik SK, Bhag M (eds) Vitro conservation and cryopreservation of tropical fruit species. IPGRI Office for South Asia and NBPGR, New Delhi, pp 191–200

Datta KB, Kanjilal B, Sarker DD (1999) Artificial seed technology: development of a protocol in Geodorum densiflorum (Lam) Schltr. - an endangered orchid. Curr Sci 76(8):1142–1145

Daud N, Taha RM, Hasbullah NA (2008) Artificial seed production from encapsulated microshoots of Saintpaulla ionantha Wendl. (African violet). J Appl Sci 8:4662–4667

Dave A, Joshi N, Purohit SD (2004) In vitro propagation of *Chlorophytum borivilianum* using encapsulated shoot buds. Eur J Hortic Sci 69:37–42

Deng Z, Hu J, Goktepe F, Harbaugh B (2007) Assessment of genetic diversity and relationships among caladium cultivars and species using molecular markers. J Am Soc Hortic Sci 132:219–229

Faisal M, Alatar AA, Ahmad N, Anis M, Hegazy AK (2012) Assessment of genetic fidelity in *Rauvolfia serpentina* plantlets grown from synthetic (encapsulated) seeds following in vitro storage at 40°C. Molecules 17:5050–5061

Gantait S, Kundu S, Ali NM, Sahu NC (2015) Synthetic seed production of medicinal plants: a review on influence of explants, encapsulation agent and matrix. Acta Physiol Plant 37:98

Gantait S, Vijayan J, Majee A (2017) Artificial seed production of *Tylophora indica* for interim storing and swapping germplasm. Horticult Plant J 3:41–46

Gray DJ, Purohit A, Triglano RN (1991) Somatic embryogenesis and development of synthetic seed technology. Crit Rev Plant Sci 10:33–61

Hassan NS (2003) In vitro propagation of jojoba (*Simmondsia chinensis* L.) through alginate-encapsulated shoot apical and axillary buds. Int J Agric Biol 5:513–516

Hung C, Trueman S (2012) Alginate encapsulation of shoot tips and nodal segments for short-term storage and distribution of the eucalypt *Corymbia torelliana* × *C. citriodora*. Acta Physiol Plant 34:117–128

Ikhlaq M, Hafiz IA, Micheli M, Ahmad T (2010) In vitro storage of synthetic seeds: effect of different storage conditions and intervals on their conversion ability. Afr J Biotechnol 9:5712–5721

Ipekci Z, Gozukirmizi N (2003) Direct somatic embryogenesis and synthetic seed production from *Paulownia elongata*. Plant Cell Rep 22:16–24

Jung SJ, Yoon ES, Jeong JH, Choi YE (2004) Enhanced post-germinative growth of encapsulated somatic embryos of Siberian ginseng by carbohydrate addition to the encapsulation matrix. Plant Cell Rep 23:365–370

Kanji M, Yuji S (2001) A method to produce encapsulatable units for synthetic seeds in *Asparagus officinalis*. Plant Cell Tissue Org Cult 64:27–32

Katouzi SSS, Majd A, Fallahian F, Bernard F (2011) Encapsulation of shoot tips in alginate beads containing salicylic acid for cold preservation and plant regeneration in sunflower (*Helianthus annuus* L.). Aust J Crop Sci 5:1469–1474

Kavyashree R, Gayatri MC, Revanasiddaiah MH (2006) Propagation of mulberry S54 by synthetic seeds of axillary buds. Plant Cell Tissue Org Cult 84:245–249

Khor E, Ng WF, Loh CS (1998) Two-coat systems for encapsulation of Spathoglottis plicata (Orchidaceae) seeds and protocorms. Biotechnol Bioeng 59(5):635–639

Kitto SK, Janick J (1982) Polyox as an artificial seed coat for asexual embryos. Hortic Sci 17:488–490

Lambardi M, Benelli C, Ozudogru EA, Ozden Tokatli Y (2006) Synthetic seed technology in ornamental plants floriculture and ornamental plant biotechnology, vol II. Global Science Books, UK

Lata H, Chandra S, Khan IA, Elsohly MA (2009) Propagation through alginate encapsulation of axillary buds of *Cannabis sativa* L.—an important medicinal plant. Physiol Mol Biol Plants 15:79–86

Lata H, Chandra S, Natascha T, Khan IA, ElSohly MA (2011) Molecular analysis of genetic fidelity in *Cannabis sativa* L. plants grown from synthetic (encapsulated) seeds following in vitro storage. Biotechnol Lett 33:2503–2508

Latif Z, Nasir IA, Riazuddin S (2007) Indigenous production of synthetic seeds in *Daucus carota*. Pak J Bot 39:849–855

Lisek A, Orlikowska T (2004) In vitro storage of strawberry and raspberry in calcium-alginate beads at 4°C. Plant Cell Tissue Org Cult 78:167–172

Mandal J, Pattnaik S, Chand PK (2000) Alginate encapsulation of axillary buds of *Ocimum americanum* L. (hoary basil), *O. basilicum* L. (sweet basil), *O. gratissimum* L. (shrubby basil), and *O. sanctum* L. (sacred basil). In Vitro Cell Dev Biol Plant 36:287–292

Mathur J, Ahuja PS, Lal N, Mathur AK (1989) Propagation of *Valeriana wallichii* DC. Using encapsulated apical and axial shoot buds. Plant Sci 60:111–116

Mehpara M, Mujib A, Siddiqui ZH (2012) Synthetic seed development and conversion to plantlet in *Catharanthus roseus* (L.) G. Don. Biotechnology 11:37–43

Mehpara M, Mujib A, Khusrau M (2015) Preparation and low temperature short-term storage for synthetic seeds of *Caladium bicolor*. Notulae Scientia Biologicae 7:90–95

Mohanraj R, Ananthan R, Bai VN (2009) Production and storage of synthetic seeds in Coelogyne breviscapa Lindl. Asian J Biotechnol 1:124–128

Mujib A, Ali M, Tonk D, Isah T, Zafar N (2016) Embryogenesis in ornamental monocots: plant growth regulators as signalling element. In: Mujib A (ed) Somatic embryogenesis in ornamentals and its application. Springer, New York, pp 187–201

Murashige T (1977) Plant cell and organ cultures as horticultural practices. In: Proceedings of the symposium on tissue culture for horticultural purposes, Ghent, Belgium, 6–9 September 1977

Nagananda GS, Satishchandra N, Rajath S (2011) Regeneration of encapsulated protocorm like bodies of medicinally important vulnerable orchid Flickingeria nodosa (Dalz.) Seidenf. Int J Bot 7(4):310

Nandini BP, Sudarshana MS, Rajashekar N (2014) Plant regeneration through somatic embryogenesis and synthetic seed production in *Rumex vesicarius* L.–a potent medicinal herb. IOSR-J Pharm Biol Sci 9:129–136

Nhut DT, Duy N, Havy NN, Khue CD, Khiem DV, Hang NTT, Vinh DN (2004) Artificial seeds for propagation of Anthurium 'TROPICAL'. J Agric Sci Technol 4:73–78

Nieves N, Lorenzo JC, Blanco MA, Gonzalez J, Peralta H, Hernandez M, Santos R, Concepcion O, Borroto CG, Borroto E, Tapia R, Martinez ME, Fundora Z, Gonzalez A (1998) Artificial endosperm of *Cleopatra tangerine* zygotic embryos: a model for somatic embryo encapsulation. Plant Cell Tissue Org Cult 54:77–83

Ozden TY, Carlo AD, Gumusel F, Pigmattelli S, Lambardi M (2008) Development of encapsulation techniques for the production and conservation of synthetic seeds in ornamental plants. Prop Ornam Plant 8:17–22

Palanyandy SR, Gantait S, Suranthran P, Sinniah UR, Subramaiam S, Aziz MA, Sarifa SRSA, Roowi SH (2015) Storage of encapsulated oil palm polyembryoids: influence of temperature and duration. In Vitro Cell Dev Biol Plant 5:118–124

Pandey A, Chand S (2005) Efficient plant regeneration from encapsulated somatic embryos of *Hyoscyamus muticus* L. Indian J Biotechnol 4:546–550

Piccioni E, Standardi A (1995) Encapsulation of micropropagated buds of six woody species. Plant Cell Tissue Organ Cult 42:221–226

Rady MR, Hanafy MS (2004) Synthetic seed technology for encapsulation and regrowth of in vitro-derived Gypsophila paniculata L. shoot tips. Arab J Biotechnol 7:251–264

Rai MK, Asthana P, Singh SK, Jaiswal VS, Jaiswal U (2009) The encapsulation technology in fruit plants-a review. Biotechnol Adv 27:671–679

Ramesh M, Marx R, Mathan G, Pandian SK (2009) Effect of bavistin on in vitro plant conversion from encapsulated uninodal microcuttings of micropropagated *Bacopa monnieri* (L)—an Ayurvedic herb. J Environ Biol 30:441–444

Ray A, Bhattacharya S (2008) Storage and plant regeneration from encapsulated shoot tips of *Rauvolfia serpentina*—an effective way of conservation and mass propagation. S Afr J Bot 74:776–779

Reddy MC, Murthy KSR, Pullaiah T (2012) Synthetic seeds: a review in agriculture and forestry. Afr J Biotechnol 11:14254–14275

Redenbaugh K (1983) Synthetic seeds. Application of synthetic seeds to crop improvement. CRC Press, Boca Raton, FL, p 481

Redenbaugh K, Nichol J, Kossler ME, Paasch B (1984) Encapsulation of somatic embryos for artificial seed production. In Vitro Cell Dev Biol Plant 20:256–257

Redenbaugh K, Fujii JA, Slade D (1988) Encapsulated plant embryos. In: Mizrahi A (ed) Advances in biotechnological processes. Alan R. Liss Inc., New York, NY, USA, pp 225–248

Refouvelet E, Nours SL, Tallon C, Daguin F (1998) A new method for in vitro propagation of lilac (Syringa vulgaris L.): regrowth and storage conditions for axillary buds encapsulated in alginate beads, development of a pre-acclimatisation stage. Sci Hortic 74:233–241

Reinert J (1958) Morphogenese und ihre kontrolle an gewebekulturen ausc arotten. Naturwissenschaften 45:344–345

Rihan HZ, Al-Issawi M, Al-Swedi F, Fuller MP (2012) The effect of using PPM (plant preservative mixture) on the development of cauliflower microshoots and the quality of artificial seed produced. Sci Hortic 141:47–52

Ruby YS, Amutha T, Priya K, Fathima MB, Manimekalai V (2015) In vitro seed germination, somatic embryogenesis and protocorm based micropropagation of a terrestrial ornamental orchid—*Spathoglottis plicata* blume. Eur J Biotechnol Biosci 3:20–23

Ruffoni B, Massabo F, Giovannini A (1994) Artificial seed technology in ornamental plants, Lasianthus and Genista. Acta Hortic 362:297–304

Sarmah DK, Borthakur M, Borua PK (2010) Artificial seed production from encapsulated PLBs regenerated from leaf base of *Vanda coerulea* Grifft. Ex Lindl. An endangered orchid. Curr Sci 98:686–690

Senaratna T (1992) Artificial seeds. Biotechnol Adv 10:379–392

Sicurani M, Piccioni E, Standardi A (2001) Micropropagation and preparation of synthetic seed in M. 26 apple stock I: attempts towards saving labour in the production of adventitious shoot tips suitable for encapsulation. Plant Cell Tissue Org Cult 66:207–216

Singh AK, Varshney R, Sharma M, Agarwal SS (2006) Bansai KC (2006) regeneration of plants from alginate encapsulated shoot tips of *Withania somnifera* (L.) Dunal.—a medicinally important plant species. J Plant Physiol 163:220–223

Singh B, Sharma S, Rani G (2007) *In vitro* response of encapsulated and non-encapsulated somatic embryos of Kinnow mandarin (*Citrus nobilis* Lour x C. *deliciosa* Tenora). Plant Biotechnol Rep 1:101–107

Srivastava V, Khan SA, Banerjee S (2009) An evaluation of genetic fidelity of encapsulated microshoots of the medicinal plant: *Cineraria maritime* following six months of storage. Plant Cell Tissue Org Cult 99:193–198

Standardi A, Micheli M, Piccioni E (1995) Incapsulamento in alginato di espianti micropropagati. Italus Hortus 2:46–52

Stephen R, Jayabalan N (2000) Artificial seed production in coriander (*Coriandrum sativum* L.). Plant Tissue Culture 10:45–49

Steward FC, Mapes MO, Smith J (1958) Growth and organized development of cultured cells. II Organization in cultures grown from freely suspended cells. Am J Bot 45:705–708

Sundararaj SG, Agrawal A, Tyagi RK (2010) Encapsulation for in vitro short-term storage and exchange of ginger (*Zingiber officinale* Rosc.) germplasm. Sci Hortic 125:761–766

Tabassum B, Nasil IA, Farooq AM, Latif Z, Husmain T (2010) Viability assessment of in vitro produced synthetic seeds of cucumber. Afr J Biotechnol 9:7026–7032

Verma SK, Rai MK, Asthana P, Jaiswal VS, Jaiswal U (2010) In vitro plantlets from alginate-encapsulated shoot tips of *Solanum nigrum* L. Sci Hortic 124:517–521

Zhang M, Peng S, Yang X, Xu L (2009) Preparation technique of Cremastra appendiculata synthetic seed. Zhongguo Zhong Yao Za Zhi 34(15):1894–1897

Zych M, Furmanowa M, Krajewska-Patan A, Lowicka A, Dreger M, Mendlewska S (2005) Micropropagation of *Rhodiola kirilowa* plants using encapsulated axillary buds and callus. Acta Biol Cracov Ser Bot 47:83–87

Cytokinin Influence on Micropropagation System of *Dianthus caryophyllus* L. 3

Iram Siddique and Vikas Yadav

Abstract

A potent, quick and refined plant regeneration protocol was effectively established for *Dianthus caryophyllus* via nodal segment explants cut out from a 2-year-old vigorous mother plant. The three diverse cytokinins viz. 6-benzyladenine (BA), Kinetin (Kin) and thidiazuron (TDZ) were assessed as enhancements to Murashige and Skoog (MS) medium, the maximum frequency of multiple shoot bud induction was attained on MS medium augmented with 5.0 μM BA and 1.0 μM α-naphthalene-acetic acid (NAA). The in vitro raised shoots when subcultured on plant growth regulator (PGR) free MS basal medium extensively amplified the rate of shoot multiplication (44.8 ± 3.5) and shoot length (7.5 ± 0.29 cm) without showing any morphological and physiological features of degeneration even after fifth subculture passage. The in vitro raised microshoots were best rooted in half-strength MS medium augmented with 2.0 μM indole-3-butyric acid (IBA). The tissue culture raised plantlets with a completely developed shoot and root system were effectively transferred to the earthen pots having garden soil and grown in greenhouse condition with a 90% survival rate. Growth parameters of 2 months old in vitro raised plants were compared with in vivo grown seedlings of the same age group.

Keywords

Caryophyllaceae · *D. caryophyllus* · In vivo · Subculture · Tissue culture

I. Siddique (✉) · V. Yadav
Plant Biotechnology Laboratory, Department of Botany, Aligarh Muslim University, Aligarh, India

I. Siddique (ed.), *Propagation and Genetic Manipulation of Plants*,
https://doi.org/10.1007/978-981-15-7736-9_3

3.1 Introduction

Dianthus caryophyllus L. (Carnation) is a herbaceous perennial plant that belongs to family Caryophyllaceae growing up to 80 cm height, cultivated all over the world for its importance as cut flowers and other decoration purposes. The significance of this decorative flower is because of its beauty, variety of colours, wonderful long-lasting quality and varied collection of numerous cultivars (Ali et al. 2008; Kanwar and Kumar 2009). The *Dianthus* flowers are an excellent accent to garlands and carnation home-flowered arrangements. It has been used as an unusual food in salad decoration, in fruit salads and for condiment fruits, and so on. The flower heads of carnation are dried up and used in cosmetics and perfumes. *Dianthus* flowers are appraised to be an antidote, antispasmodic, cardiotonic, diaphoretic and nervine (Chopra et al. 1986).

In current years, there has been bigger attention to micropropagation practices that provide a practicable means for large scale proliferation and germplasm preservation of rare, endangered, aromatic and medicinal plants (Faisal et al. 2006; Siddique et al. 2010). These techniques could be a cost-effective means for mass propagation of the best planting material all over the year, without any seasonal limitations. Effective protocols have been published for the micropropagation of valuable plants from leaf and nodal segment explants (Van Altvorst et al. 1994, 1995; Kantia and Kothari 2002) and the stimulation of direct somatic embryogenesis from leaf, petal and anther explants (Pareek and Kothari 2003; Karami et al. 2006; XiaoPeng et al. 2008) in *Dianthus*.

As far as we know, very few reports are present on the successful utilization of TDZ for fast and effective mass multiplication of Carnation (Frey and Janick 1991; Ahmad et al. 2006b; Casas et al. 2010), and the existing protocol yielded a low number of shoots that cannot fulfil the demand of this plant on a large scale. Therefore, the present research work has been assumed to prove the relative efficiency of TDZ, BA and Kin with those of earlier reports for in vitro plant regeneration from nodal segment explants and to detect the optimal concentration of cytokinins exposure for maximum shoot multiplication and formation of complete plantlets.

3.2 Materials and Methods

3.2.1 Nodal Explants and Surface Sterilization

The nodal segment explants of *Dianthus* were taken from the nursery-grown 2 years old healthy mother plant. Nodal segments were fully splashed under running tap water for about 30 min and followed by soaking in 5% (v/v) solution of a liquid detergent Labolene (Qualigens, Mumbai, India) for about 5 min. Following repetitive rinses with sterile, single distilled water, the nodal explants were again surface sterilized with freshly prepared 0.1% (w/v) mercuric chloride (Qualigens, Mumbai,

India) for 5 min under laminar flow hood and followed by repetitive rinses with sterile double distilled water.

3.2.2 Culture Media and Culture Room Conditions

Murashige and Skoog (1962) medium composed with 3% (w/v) sucrose and 0.8% agar (w/v) or 0.25% (w/v) Gelrite was used during the experiment. Plant growth regulators (PGRs) viz. BA, Kin and TDZ either singly or in combinations with auxin (NAA) were supplemented to the MS medium as stated later. The pH of the medium was adjusted to 5.8 using 1 N NaOH or 1 N HCl earlier to autoclaving at 121 °C for 15 min. All the cultures were placed in culture room at 25 ± 2 °C under a 16/8 h light/dark control, under a photon flux density of 50 μ mol m^{-2} s^{-1} provided by 40 W white fluorescent tubes and at 55 ± 5% relative humidity.

3.2.3 Effect of PGRs on Shoot Induction and Multiplication

For induction of multiple shoot regeneration, nodal segments having axillary shoot meristem were inoculated on MS medium augmented with BA, Kin and TDZ at 0.5, 2.5, 5.0, 7.5 and 10.0 μM either singly or in combination with NAA (0.1, 0.5 and 1.0 μM). Data for shoot bud regeneration and multiplication were observed after 4 and 8 weeks of incubation.

3.2.4 In Vitro Rooting in Microshoots

Well-developed healthy microshoots (3–4 cm long) were removed and transferred on to rooting medium comprised of full and half-strength MS medium supplemented with auxin (IBA) at diverse concentrations (0.5, 1.0, 1.5 and 2.0 μM). Data were noted as the percentage response of rooting, the mean number of roots/shoots with their average length after 4 weeks of incubation.

3.2.5 Hardening and Acclimatization of Plantlets

In vitro raised plantlets with the healthy developed shoot (5–6 cm) and roots (4–5 cm) were taken out from the culture tubes and washed carefully under running tap water to remove all the traces of agar from the roots. Then, they were transferred to thermocol cups holding disinfected Soilrite and retained under diffuse light (16/8 h photoperiod) conditions. Potted plants were covered with transparent polythene bags to confirm high levels of humidity around the plants and irrigated every third day with half-strength MS salt solution up to 2 weeks. Polythene bags were removed above the cups after 2 weeks in order to acclimatize the plants to natural field conditions. After 4 weeks, completely acclimatized plants were transferred to earthen pots having normal garden soil and kept in a greenhouse under normal day length conditions.

3.2.6 Statistical Analysis

All the experiments were carried out with 20 replicates with three repetitions each. The whole data were analysed statistically by using SPSS Ver. 14 (SPSS Inc., Chicago, USA). The significance of variances among means was accredited using Duncan's multiple range test at $P = 0.05$ and data denoted as mean ± standard error (SE).

3.3 Result and Discussion

3.3.1 Induction and Multiplication Phase

The in vitro morphogenetic results of nodal segment explants to different cytokinins tested (BA, Kin and TDZ), either singly or in combination with an auxin (NAA), are outlined in Tables 3.1 and 3.2. No axillary bud shoot induction was detected when nodal segment explants were inoculated on PGR-free MS basal medium. Although a distinctive response with respect to axillary shoot bud induction was detected on

Table 3.1 Effect of plant growth regulators on shoot regeneration from nodal explants of *D. caryophyllus* in MS Medium after 4 weeks of culture

Plant growth regulators (μM)			% Regeneration	Mean number of shoots/explant	Mean shoot length (cm)
BA	Kin	TDZ			
0.0	0.0	0.0	0.0	0.0^{j}	0.0^{g}
0.5			60	$1.9 \pm 0.31^{h,i}$	$1.9 \pm 0.31^{d,e,f}$
2.5			69	3.1 ± 0.34^{g}	$2.6 \pm 0.29^{c,d}$
5.0			80	5.3 ± 0.26^{e}	3.7 ± 0.29^{b}
7.5			71	$3.8 \pm 0.26^{f,g}$	$2.8 \pm 0.37^{c,d}$
10.0			64	$1.4 \pm 0.23^{h,i}$	1.5 ± 0.20^{f}
	0.5		55	$1.3 \pm 0.20^{h,i}$	$1.6 \pm 0.35^{e,f}$
	2.5		62	2.1 ± 0.23^{h}	$3.0 \pm 0.23^{b,c}$
	5.0		71	$3.6 \pm 0.35^{f,g}$	3.8 ± 0.25^{b}
	7.5		63	$1.9 \pm 0.17^{h,i}$	$2.7 \pm 0.23^{c,d}$
	10.0		56	1.0 ± 0.26^{i}	1.2 ± 0.08^{f}
		0.5	68	2.3 ± 0.29^{f}	$2.4 \pm 0.26^{d,e}$
		2.5	77	3.5 ± 0.40^{d}	$3.5 \pm 0.23^{b,c}$
		5.0	88	7.0 ± 0.89^{a}	4.1 ± 0.23^{a}
		7.5	70	6.2 ± 0.37^{b}	3.5 ± 0.35^{b}
		10.0	61	1.6 ± 0.49^{c}	2.2 ± 0.20^{b}

Values represent means ± SE. Means followed by the same letter within columns are not significantly different ($P = 0.05$) using Duncan's multiple range test

Table 3.2 Effect of different concentrations of NAA with an optimal concentration of BA and Kin (5.0 μM) on shoot multiplication and elongation from nodal explants of *D. caryophyllus* in MS Medium after 8 weeks of culture

Plant growth regulators (μM)	Mean number of shoots/explant	Mean shoot length (cm)
BA (5.0) + NAA (0.1)	5.9 ± 0.20[c]	4.4 ± 0.29[b,c]
BA (5.0) + NAA (0.5)	8.7 ± 0.43[b]	5.6 ± 0.20[b]
BA (5.0) + NAA (1.0)	6.9 ± 0.53[a]	3.8 ± 0.32[a]
Kin (5.0) + NAA (0.1)	4.5 ± 0.55[d]	4.0 ± 0.17[c]
Kin (5.0) + NAA (0.5)	7.6 ± 0.60[c]	4.8 ± 0.35[c]
Kin (5.0) + NAA (1.0)	5.9 ± 0.71[c]	3.5 ± 0.35[b,c]

Values represent means ± SE. Means followed by the same letter within columns are not significantly different ($P = 0.05$) using Duncan's multiple range test

medium augmented with a varying range of cytokinins viz. BA, TDZ and Kin (0.5–10 μM). TDZ was found more potent than other cytokinins used and a concentration of 5.0 μM was proved to be optimum for the induction of maximum percentage regeneration response (88%), the mean number of shoots/explant (7.0 ± 0.89) and average shoot length (4.1 ± 0.23 cm) (Fig. 3.2a, b). The effect of TDZ on multiple shoot regeneration has been reported in previous studies of many different ornamental plant species (Nontaswatsri et al. 2002; Sivanesan et al. 2011). The percentage response of multiple shoot regeneration, the mean number of shoots/explant and their average shoot length were gradually increased with an increase in the concentration of cytokinin from 0.5 to 5.0 μM. MS medium comprising BA induced 5.3 ± 0.26 mean shoots/explants in 80% cultures, while Kin at the similar concentration regenerated 3.6 ± 0.35 mean shoots/explant with 71% regeneration frequency (Table 3.1). All in vitro regenerated microshoots were free from basal callusing at their cut ends. The percentage of shoot regeneration frequency and the mean number of shoots/explants with their average shoot length was dropped immediately with a rise in the concentration of all the cytokinin tested beyond the optimal level (5.0 μM).

To make more improvements in shoot multiplication frequency, NAA at diverse concentrations (0.1, 0.5 and 1.0 μM) was tried with MS medium augmented with optimal cytokinin concentration (5.0 μM) of BA and Kin (Table 3.2). The combination of NAA with cytokinin triggered the regeneration of multiple shoots having healthy leaves during subculture. Out of all the cytokinin–auxin combination treatments tested, the maximum number of mean shoots/explant (8.7 ± 0.43) with an average shoot length (5.6 ± 0.20 cm) was attained at the MS + 5.0 μM BA +0.5 μM NAA. Therefore, this growth regulator combination treatment was considered as optimal for maximum shoot regeneration and multiplication among all the BA, Kin and NAA combination treatments examined. The outcomes found here exhibited reliability with other findings where BA and NAA stimulated the regeneration and proliferation of shoots in several plant species (Miller et al. 1991; Kantia and Kothari 2002; Remya et al. 2013).

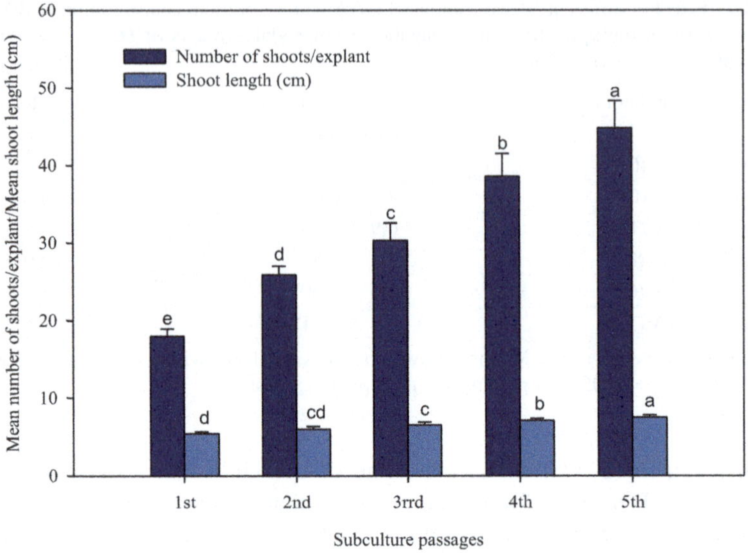

Fig. 3.1 The evaluation of the morphogenetic potential of shoot culture obtained from TDZ (5.0 μM) after being tested for five subculture passages on growth regulator free MS medium. Bars represent mean ± SE. Means followed by the same letter are not significantly different ($P = 0.05$) using Duncan's multiple range test

Shoot buds regenerated from nodal segment explants on TDZ supplemented MS medium did not lengthen properly and resulted in the formation of a rosette of regenerated shoots when continuously transferred on the similar media composition. TDZ generally induced shoot multiplication and obstructed their elongation, and hence, might be dependable with its intense cytokinin action. The difficulty of shoot elongation was solved by a subculture of aggregate shoot mass to another medium devoid of TDZ, where the further elongation of multiple shoots was taking place within 4 weeks of subculture.

The influence of different subculture passages was also assessed by proliferation and elongation of shoots on TDZ supplemented MS medium after transfer to MS basal medium free from any TDZ concentration (Fig. 3.1). The in vitro cultures that had grown-up constantly on TDZ comprising media showed fasciation and distortion of the apical shoot meristem. The damaging effect of the persistent occurrence of TDZ on the multiplication and elongation has been testified in a number of plant species (Ahmad et al. 2006a; Siddique et al. 2013). In this research analysis, the mean number of shoots/explant (44.8 ± 3.5) and average shoot length (7.5 ± 0.29 cm) (Fig. 3.2c, d) continued to proliferate up to fifth subculture passages without any mark of degeneration. A related trend of subculture passages was also reported by Tiwari et al. (2001) and Siddique and Anis (2007).

Fig. 3.2 (**a–f**) In vitro shoot multiplication and plantlet formation in *D. caryophyllus*. (**a**, **b**) Induction of multiple shoots on MS + TDZ (5.0 μM). (**c**, **d**) Shoot multiplication and elongation on MS medium devoid of TDZ. (**e**) In vitro rooting on ½ MS + IBA (2.0 μM). (**f**) In vitro flowering in an acclimatized micropropagated plant

3.3.2 Rooting of Microshoots

Healthy developed microshoots (3–4 cm) were taken out from in vitro raised cultures and placed into the full and half-strength MS medium supplemented with a diverse concentration of IBA (0.5, 1.0, 1.5 and 2.0 μM) (Table 3.3). Half-strength MS

Table 3.3 Effect of MS strength and IBA concentration on root induction from in vitro raised shoots of *D. caryophyllus* after 4 weeks of culture

Treatment	% Rooting	Mean no. of roots/ shoot	Mean root length (cm)
MS	60	1.8 ± 0.11^{d}	1.2 ± 0.14^{d}
(1/2)MS	65	$3.1 \pm 0.14^{b,c}$	2.0 ± 0.12^{c}
(1/2)MS + IBA (0.5 μM)	70	3.5 ± 0.17^{b}	2.8 ± 0.17^{b}
(1/2)MS + IBA (1.0 μM)	75	4.0 ± 0.23^{a}	3.0 ± 0.29^{a}
(1/2)MS + IBA (1.5 μM)	79	4.2 ± 0.27^{b}	3.6 ± 0.17^{a}
(1/2)MS + IBA (2.0 μM)	83	5.4 ± 0.20^{c}	4.0 ± 0.20^{b}

Values represent means ± SE. Means followed by the same letter within columns are not significantly different ($P = 0.05$) using Duncan's multiple range test

medium was better than full strength MS medium and half-strength MS medium augmented with 2.0 μM IBA induced the production of 5.4 mean roots/shoot in 83% cultures after 4 weeks of transfer in rooting medium (Fig. 3.2e). The efficacy of IBA for induction of in vitro rooting has been recorded by Siddique and Anis 2007; Anis et al. 2010.

3.3.3 Acclimatization

In vitro raised plants having 4–6 completely extended leaves and fully grown root system were effectively acclimatized in the culture room conditions by planting them in disinfected Soilrite for 4 weeks and were finally transferred into the garden soil (Fig. 3.2f). About 90% of the in vitro raised plants survived when transferred from Soilrite to garden soil and did not show any noticeable dissimilarity with respect to plant morphology and other growth parameters.

References

Ahmad N, Siddique I, Anis M (2006a) Improved plant regeneration in *Capsicum annuum* L. from nodal segments. Biol Plant 50:701–704

Ahmad N, Srivastava R, Anis M (2006b) Improvement in carnation shoot multiplication using thidiazuron in vitro. Propag Ornament Plant 6:109–113

Ali A, Afrasiab H, Naz Rauf M, Iqbal J (2008) An efficient protocol for in vitro propagation of carnation (*Dianthus caryophyllus*). Pak J Bot 40:111–121

Anis M, Varshney A, Siddique I (2010) In vitro clonal propagation of *Balanites aegyptiaca* (L.) Del. Agrofor Syst 78:151–158

Casas J, Olmas E, Piqueras A (2010) In vitro propagation of carnation (*Dianthus caryophyllus*). In: Jain SM, Ochatt SJ (eds) Protocols for in vitro propagation of ornamental plants, Springer Protocols. Humana Press, New York, USA, pp 109–116

Chopra RN, Mayar SL, Chopra IC (1986) Glossary of Indian medicinal plants (including the supplement). Council of Scientific and Industrial Research (CSIR), New Delhi, India

Faisal M, Siddique I, Anis M (2006) An efficient plant regeneration system for *Mucuna pruriens* L. (DC.) using cotyledonary node explants. In Vitro Cell Develop Biol Plant 42:59–64

Frey L, Janick J (1991) Organogenesis in carnation. J Am Soc Hortic Sci 116:1108–1112
Kantia A, Kothari SL (2002) High frequency adventitious shoot bud formation and plant regeneration from leaf explants of *Dianthus chinensis* L. Sci Hortic 96:205–212
Kanwar JK, Kumar S (2009) Influence of growth regulators and explants on shoot regeneration in carnation. Hortic Sci 36:140–146
Karami O, Deljou A, Esna-Ashari M, Ostad-Ahmadi P (2006) Effect of sucrose concentrations on somatic embryogenesis in carnation (*Dianthus caryophyllus* L.). Sci Hortic 110:340–344
Miller MR, Kaul V, Hutchinson FJ, Maheswaran G, Richards D (1991) Shoot regeneration from flower buds of carnation (*Dianthus caryophyllus*). Ann Bot 68:563–568
Murashige T, Skoog F (1962) A revised medium for rapid growth and bioassays with tobacco tissue cultures. Physiol Plant 15:473–497
Nontaswatsri C, Fukai S, Touma T, Goi M (2002) Comparison of adventitious shoot formation from node and leaf explants of various carnations (*Dianthus caryophyllus* L.) cultivars. J Hortic Sci Biotechnol 77:520–525
Pareek A, Kothari SL (2003) Direct somatic embryogenesis and plant regeneration from leaf cultures of ornamental species of *Dianthus*. Sci Hortic 98:449–459
Remya M, Bai VN, Mutharaian VN (2013) In vitro regeneration of *Aristolochia tagala* and production of artificial seeds. Biol Plant 57:210–218
Siddique I, Anis M (2007) Rapid micropropagation of *Ocimum basilicum* using shoot tip explants pre-cultured in thidiazuron supplemented liquid medium. Biol Plant 51:787–790
Siddique I, Anis M, Aref IM (2010) In vitro adventitious shoot regeneration via indirect organogenesis from petiole explants of *Cassia angustifolia* Vahl.—a potential medicinal plant. Appl Biochem Biotechnol 162:2067–2074
Siddique I, Bukhari NA, Perveen K, Siddiqui I, Anis M (2013) Pre-culturing of nodal explants in thidiazuron supplemented liquid medium improves in vitro shoot multiplication of *Cassia angustifolia*. Acta Biol Hung 64:377–384
Sivanesan I, Song JY, Hwang SJ, Jeong BR (2011) Micropropagation of *Cotoneaster wilsonii* Nakai-a rare endemic ornamental plant. Plant Cell Tissue Org Cult 105:55–63
Tiwari V, Tiwari KN, Singh BD (2001) Comparative studies of cytokinins on in vitro propagation of *Bacopa monniera*. Plant Cell Tissue Org Cult 66:9–16
Van Altvorst AC, Koehorst HJJ, Bruinsna T, Dons JJM (1994) Improvement of adventitious shoot formation from carnation leaf explants. Plant Cell Tissue Org Cult 37:514–518
Van Altvorst AC, Yancheva S, Dons H (1995) Cell within the nodal region of carnation shoots exhibits a high potential for adventitious shoot formation. Plant Cell Tissue Org Cult 40:151–157
Xiaopeng F, Shul Hua Y, Man Zhu B (2008) Factors affecting somatic embryogenesis in anther cultures of Chinese pink (*Dianthus chinensis* L.). In Vitro Cell Develop Biol Plants 44:194–202

Phenolic Compound Contents of *Hypericum* Species from Turkey

4

Hilal Surmuş Asan

Abstract

Hypericaceae is a small cosmopolite dicot family of tree, shrub, and herb. *Hypericum* is the largest genus of Hypericaceae comprising of about 484 species throughout the world. In Turkey, *Hypericum* is represented by 96 species, 104 taxa in which 45 of them are endemic. The genus *Hypericum* is known worldwide with its wide folk medicinal usage. The extracts of *Hypericum perforatum* L., the most abundant and well-known species, are now widely used in Europe as a drug for the treatment of mild-to-moderate depression. In Turkish traditional medicine, *Hypericum* L. species have been known under the names "kantaron, peygamber çiçeği, kılıçotu, kanotu, kuzukıran, and binbirdelik otu" and used for the treatment of burns, wounds, bacterial and viral infections, hemorrhoids, diarrhea, and ulcers. The pharmacological studies revealed that this species has antitumor, antidepressant, anti-inflammatory, anti-microbial, anti-viral, anti-nociceptive, and wound healing activity. The value of *Hypericum* plants has been mainly attributed to the phytochemicals, namely naphthodianthrones (e.g., hypericin and pseudohypericin), acylphloroglucinol derivatives (e.g., hyperforin and adhyperforin), flavonoids (e.g., quercetin, quercitrin, hyperoside, and biapigenin), *n*-alkanes, tannins, xanthones and essential oils they possess.

Due to their phytochemical composition and waste usage in traditional folk medicine, many studies were made on biological activities of different species extracts of *Hypericum* in Turkey and the world. Turkey is an important prevalence center of *Hypericum* species. This review will investigate the phenolic compound contents of *Hypericum* species in Turkish flora.

H. S. Asan (✉)
Science Faculty, Biology Department, Dicle University, Diyarbakır, Turkey

I. Siddique (ed.), *Propagation and Genetic Manipulation of Plants*,
https://doi.org/10.1007/978-981-15-7736-9_4

Keywords

Hypericum · Phenolic compound · Hypericins · Turkey

4.1 Introduction

The genus *Hypericum* (Hypericaceae), includes 484 species, is lived mainly in varied temperate regions all over the world, and is found in forms of small trees, shrubs, and herbs. Recently, 36 taxonomic sections have been recognized within *Hypericum* genus (Crockett and Robson 2011).

Scientific studies on *Hypericum* species are concentrated as its members have yielded many bioactive compounds namely, hyperforin and adhyperforin (phloroglucinol derivatives), hypericin and pseudohypericin (naphthodianthrones), hyperoside, rutin, quercitrin, quercetin, and biapigenin (flavonoids), caffeic and chlorogenic acids (the phenylpropanes) and essential oils that show many biological activities (Kasper et al. 2010). *Hypericum* species have been used as healing agents due to their variety of medicinal properties for centuries. The best known of one is *H. perforatum* L. and extracts of *H. perforatum* L., the most prevalent and domesticated species, are now largely used in Europe as a drug for the treatment of depression (Fiebich et al. 2011). Turkey is an important geographical distribution center of *Hypericum* species. According to the most recent revision of *Hypericum* genus carried out by Guner et al. (2012), there are a total of 96 *Hypericum* species in the flora of Turkey, 46 of which are endemic. *Hypericum* species have been used traditionally as sedatives, antiseptics, and antispasmodics in Turkey. They are known under several names such as "kantaron, peygamber çiçeği, kılıçotu, kanotu, kuzukıran, and binbirdelik otu" (Bingol et al. 2011). The aim of the present study was to review and summarize noteworthy studies on the phenolic contents of Turkish *Hypericum species.*

4.1.1 The Floristic Characters of *Hypericum* L.

The floristic characters of *Hypericum* genus can be defined as (The Flora of Turkey–Vol 2) "shrubs or herbs are usually with translucent glands containing essential oils and sometimes red or black glands containing hypericin. Leaves simple, opposite or rarely whorled. Flowers bisexual, Sepals 5, Petals 5, usually yellow, often tinged red or with red veins, rarely with nectary appendages. Stamens in 5 fascicles, anti-petalous, free or 4 of them united in pairs to form 2 anti-sepalous compound fascicles, with 3-c. 125 stamens in each, rarely alternating with sterile fascicles ('fasciclodes'). Ovary superior with axile or parietal placentation, 3-5-locular or partly or completely 1-locular with 2-many ovules on each placenta. Styles 3-5, free, slender, Fruit capsular, dehiscing septicidally, usually with resin-containing vittae or vesicles in the wall, or rarely fleshy and indehiscent. The distribution and configuration of glands is important in classification. In the following account, glands are described as marginal where they interrupt the contour of the organ, intramarginal where they occur near the margin without interrupting the contour and superficial where they occur away from the margin. Narrow glands in the ovary and capsule walls are termed *vittae*, and short swollen glands are described as *vesicles*" Davis et al. 1967).

4.1.2 *Hyperiicum* Species in Turkey

Turkey is an important geographical distribution center of *Hypericum* species. According to the most recent revision of *Hypericum* genus carried out by Guner et al. (2012), there are a total of 96 *Hypericum* species in the flora of Turkey, 46 of which are endemic (Table 4.1).

4.1.3 Traditional Uses of *Hypericum* Species in Turkey

Turkey is an important center for *Hypericum* species, which are known as "kantaron, binbirdelik otu, kan otu, kılıç otu, yaraotu, kuzukıran" (Bingol et al. 2011). The different parts of these plant are used in several forms like ointment, decoction, and oleate (dried herbs are kept in olive oil for 1 month and filtered). These species have been used traditionally in Anatolia for the treatment of burns, skin injuries, (bacterial and viral infections), hemorrhoids, diarrhea, and ulcers for centuries. The species that have traditional uses are being shown below (Table 4.2).

4.2 Phenolic Compounds of *Hypericum* Species

Hypericum species have many important chemical compounds possessing biological activities. These compounds are hypericin and pseudohypericin (naphthodianthrones), hyperforin and adhyperforin (acylphloroglucinol derivatives), quercetin, quercitrin, hyperoside and biapigenin (flavonoids) (Bombardelli and Morazzoni 1995; Bruneton 1995), especially 3,8-biapigenin and amentoflavone (biflavones) (Berghoefer and Hoelzl 1987, 1989), chlorogenic acid, p-coumaric acid and caffeic acid (phenylpropanes) (Nahrstedt and Butterweck 1997), and lesser amounts of tannins and xanthones.

4.2.1 Naphthodianthrones

Naphthodianthrones are typical compounds of the genus *Hypericum* (Hegnauer 1989). These compounds have an intense red color and phototoxic properties (Schey et al. 2000). Hypericins are the major components of this group of compounds. These two compounds are called hypericin (Fig. 4.1) and

Table 4.1 *Hypericum* species in the flora of Turkey

	Hypericum species	Endemic
1	*H. adenotrichum* Spach	+
2	*H. amblysepalum* Hochst.	
3	*H. androsaemum* L.	
4	*H. apricum* Kar. and Kir.	
5	*H. armenum* Jaub. and Spach	
6	*H. atomarium* Boiss.	
7	*H. aucheri* Jaub. and Spach	
8	*H. avicularifolium* Jaub. and Spach subsp. *avicularifolium* var. *albiflorum* Hub.-Mor.	+
9	*H. avicularifolium* Jaub. and Spach subsp. *avicularifolium* var. *avicularifolium*	+
10	*H. avicularifolium* Jaub. and Spach subsp. *byzantinum* (Azn.) Robson	+
11	*H. avicularifolium* Jaub. and Spach subsp. *depilatum* (Freyn and Bornm.) Robson var.	+
12	*bourgaei* (Boiss.) Robson	
13	*H. avicularifolium* Jaub. and Spach subsp. *depilatum* (Freyn and Bornm.) Robson var.	+
14	depilatum	
15	*H. avicularifolium* Jaub. and Spach subsp. *depilatum* (Freyn and Bornm.) Robson var.	+
16	*leprosum* (Boiss.) Robson	
17	*H. avicularifolium* Jaub. and Spach subsp. *uniflorum* (Boiss. and Heldr.) Robson	+
18	*H. bithynicum* Boiss.	
19	*H. bupleuroides* Gris.	
20	*H. calycinum* L.	
21	*H. capitatum* Choisy var. *capitatum*	+
22	*H. capitatum* Choisy var. *luteum* Robson	
23	*H. cardiophyllum* Boiss.	+
24	*H. cerastoides* (Spach) Robson	
25	*H. confertum* Choisy subsp. *confertum*	+
26	*H. confertum* Choisy subsp. *stenobotrys* (Boiss.) Holmboe	
27	*H. crenulatum* Boiss.	+
28	*H. cuisinii* Barbey	
29	*H. davisii* Robson	
30	*H. elegans* Steph. ex Willd.	
31	*H. elongatum* Ledeb. subsp. *apiculatum* Robson	
32	*H. elongatum* Ledeb. subsp. *elongatum*	
33	*H. elongatum* Ledeb. subsp. *microcalycinum* (Boiss. and Heldr.) Robson	
34	*H. empetrifolium* Willd.	
35	*H. fissurale* Woron	+
36	*H. formosissimum* Takht.	+
37	*H. havvae* Guner	+
38	*H. helianthemoides* (Spach) Boiss.	

(continued)

Table 4.1 (continued)

	Hypericum species	Endemic
39	*H. heterophyllum* ~~VV~~ent.	+
40	*H. hircinum* L. subsp. *albimontanum* (Greuter) Robson	
41	*H. hircinum* L. subsp. *majus* (Aiton) Robson	
42	*H. hirsutum* L.	
43	*H. huber-morathii* Robson	+
44	*H. imbricatum* Poulter	+
45	*H. kazdagensis* Gemici and Leblebici	+
46	*H. kotschyanum* Boiss.	+
47	*H. lanuginosum* Lam. var. *lanuginosum*	
48	*H. lanuginosum* Lam. var. *pestalozzae* (Boiss.) Robson	+
49	*H. lanuginosum* Lam. var. *scabrellum* (Boiss.) Robson	+
50	*H. linarioides* Bosse	
51	*H. lydium* Boiss.	
52	*H. lysimachioides* Boiss. and Noë var. *lysimachioides*	
53	*H. lysimachioides* Boiss. and Noë var. *spathulatum* Robson	
54	*H. malatyanum* Peçmen	+
55	*H. marginatum* Woron	+
56	*H. minutum* Davis and Poulter	+
57	*H. monadenum* Robson apud Poulter	+
58	*H. montanum* L.	
59	*H. montbretii* Spach	
60	*H. neurocalycinum* Boiss. and Heldr.	+
61	*H. nummularioides* Trautv.	+
62	*H. olivieri* (Spach) Boiss.	
63	*H. olympicum* L. subsp. *olympicum*	
64	*H. olympicum* L. subsp. *macrocalyx* (Freyn) Robson	+
65	*H. orientale* L.	
66	*H. origanifolium* Willd.	
67	*H. pallens* Banks and Sol	
68	*H. pamphylicum* Robson and Davis	+
69	*H. perfoliatum* L.	
70	*H. perforatum* L.	
71	*H. peshmenii* Yıldırımlı	+
72	*H. polyphyllum* Boiss. and Bal. subsp. *lycium* Robson and Hub.-Mor.	+
73	*H. polyphyllum* Boiss. and Bal. subsp. *polyphyllum*	+
74	*H. polyphyllum* Boiss. and Bal. subsp. *subcordatum* Robson and Hub.-Mor.	+
75	*H. pruinatum* Boiss. and Bal.	
76	*H. pseudolaeve* Robson	+
77	*H. pumilio* Bornm.	+
78	*H. retusum* Aucher	
79	*H. rumeliacum* Boiss. subsp. *rumeliacum*	+
80	*H. rupestre* Jaub. and Spach	+
81	*H. russeggeri* (Fenzl) R. Keller	+

(continued)

Table 4.1 (continued)

	Hypericum species	Endemic
82	*H. salsolifolium* Hand.-Mazz	+
83	*H. salsugineum* Robson and Hub.-Mor.	+
84	*H. saxifragum* Robson and Hub.-Mor.	+
85	*H. saxifragum R*obson and Hub.-Mor. subsp. *eglandulosum* Parolly and Eren	
86	*H. scabroides* Robson and Poulter	+
87	*H. scabrum* L.	
88	*H. sorgerae* Robson	+
89	*H. spectabile* Jaub. and Spach	+
90	*H. ternatum* Poulter	+
91	*H. tetrapterum* Fries	
92	*H. thasium* Griseb.	+
93	*H. thymbrifolium* Boiss. and Noë	+
94	*H. thymifolium* Banks and Sol.	
95	*H. thymopsis* Boiss.	+
96	*H. triquetrifolium* Turra	
97	*H. uniglandulosum* Hausskn. ex Bornm	+
98	*H. vacciniifolium* Hayek and Siehe	+
99	*H. venustum* Fenzl	
100	*H. vesiculosum* Griseb	+
101	*H. xylosteifolium* (Spach) Robson	

Davis et al. (1967); Ekim et al. (2000); Güner et al. (2000)

pseudohypericin (Fig. 4.2), respectively. Hypericin and pseudohypericin occur in the flowers and leaves of the *Hypericum* plants (Cellarova et al. 1994).

4.2.2 Phloroglucinols

Phloroglucinol derivatives are widely distributed in the genus *Hypericum*. Two closely related phloroglucinol compounds are hyperforin (Fig. 4.3) and adhyperforin (Hobbs 1989). These are found only in the reproductive parts (Tekelova et al. 2000).

4.2.3 Flavonoids

The major group of biologically active compounds in *Hypericum* species comprises of flavonoids. Kaempferol, luteolin, myricetin, quercetin (Fig. 4.4), hyperoside (hyperin) (Fig. 4.5), and rutin; quercitrin; and isoquercitrin are samples of flavonoid compounds (Kurth and Spreemann 1998; Hansen et al. 1999; Pietta et al. 2001; Naeem et al. 2010).

Table 4.2 The traditional uses of *Hypericum* species in Turkey

Hypericum species	Traditional uses	References
H. atomarium	Enteritis, stomach diseases, sedative	Vural (2008)
H. aviculariifolium	Urethra diseases	Keskin and Alpınar (2002)
H. aviculariifolium subsp. depilatum	Hemorrhoid, psoriasis	Baytop (1999)
Hypericum bupleuroides Gris.	Burns, intestinal disorders	Baytop (1999)
H. cerastoides	Diarrhea	Kizilarslan (2008)
H. calycinum	Anti-asthmatic, remove spasm	Aktan (2011)
H. capitatum var. capitatum	To release pain	Akan et al. (2013)
H. confertum subsp. *confertum*	Wound healing, anti-asthmatic	Bulut (2006); Buyukgebiz (2006); Mart (2006)
H. confertum var. *stenobotrys*	Stomach diseases, remove spasm	Buyukgebiz (2006); Bulut (2006); Mart (2006)
H. heterophyllum	Anti-inflammatory	Unal et al. (2008)
H. hyssopifolium and *H. hyssopifolium* subsp. *elongatum* var. *elongatum*	Hemorrhoid, remove spasm, diarrhea, sedative, anti-helminthic, antiseptic, antifungal, psoriasis	Unal et al. (2008)
H. lydium	Stomach diseases, indigestions, hemorrhoid, wound healer, menstrual disorders	Yeşilada et al. (1995); Davis (1988); Sezik et al. (2001); Yesil (2007); Altundag and Ozturk (2011)
H. montbretii	Kidney stones, eczema, stomach disorders, ulcers, hemorrhoids	Altundag and Ozturk (2011); Keskin and Alpınar (2002)
H. olympicum	For inflamed wounds, stomach ache, cuts.	Tuzlaci and Aymaz (2001)
H. orientale	Stomach diseases, sedative, hemorrhoid, psoriasis	Ezer and Mumcu (2006); Tatli et al. (2009); Sezik et al. (2001); Baytop (1999)
H. perforatum	Anti-inflammatory, anti-hemorrhagic, antifungal, antihypertensive, kidney stones, eczema, urinary diseases, cold, cardiac diseases, arteriosclerosis, bladder ailments, facial paralysis, gastritis, chest diseases, internal hemorrhage, stomach diseases, bronchitis, tuberculosis, pharyngitis, wounds, burns, cuts, herpes labialis, lip chap, diabetes, enteritis, ulcers	Demirci (2010); Cimen (2007); Kultur (2007); Sezik et al. (2001); Tuzlaci and Aymaz (2001); Ezer and Avcı (2004); Ecevit and Ozhatay (2006); Yesilada et al. (1993); Yeşilada et al. (1995); Tuzlaci and Aymaz (2001)
H. scabrum	Hemorrhoid, antispasmodic, anti-inflammatory, sedative, constipation, peptic ulcer	Unal et al. (2008); Ezer and Mumcu (2006); Tanker (1971); Yeşilada et al. (1995)

(continued)

Table 4.2 (continued)

Hypericum species	Traditional uses	References
H. ternatum	Antiasthmatic, wound healing	Bulut (2006); Mart (2006)
H. thymifolium	Stomach diseases	Mart (2006)
H. triquetrifolium	Diabetes, cardiac diseases, burns, gastrointestinal diseases	Akgul (2008); Baytop (1999)

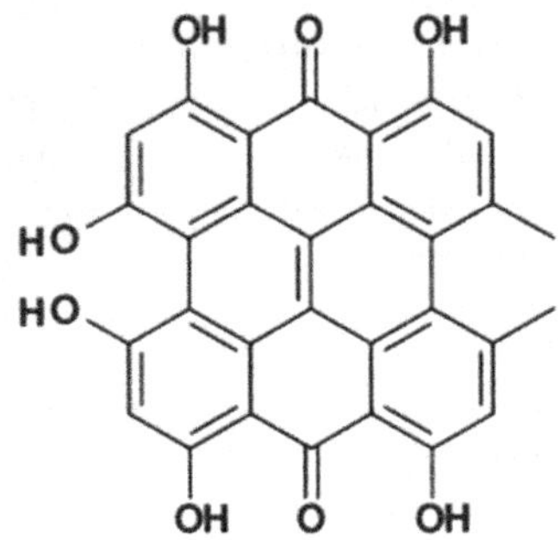

Fig. 4.1 Chemical structure of hypericin

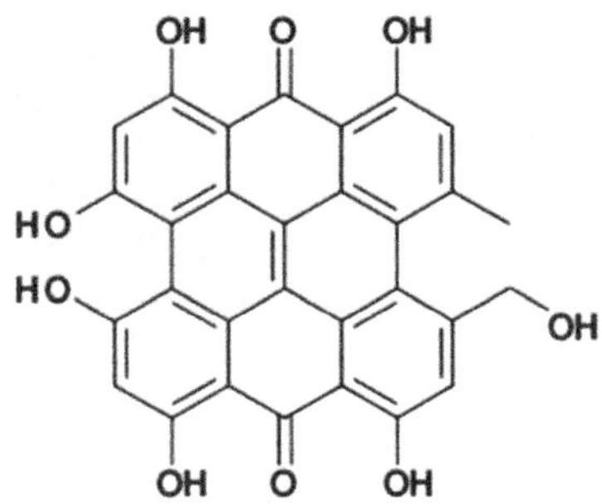

Fig. 4.2 Chemical structure of pseudohypericin

4.2.4 Biflavones

Biflavones are a rare group of flavones found in some plants. Three biflavones detected in *H. perforatum* are 3′, 8″-biapigenin, amentoflavone (Berghoefer and Hoelzl 1987, 1989), and 6′, 8″ -diquercetin (Kurkin and Pravdivtseva 2007).

4.2.5 Phenylpropanes

Caffeic, p-coumaric, ferulic, isoferulic, gentisic (Hobbs 1989; Upton 1997), shikimic acids (Bilia et al. 2001), and chlorogenic acid (Nahrstedt and Butterweck 1997) can be given as compounds in this group.

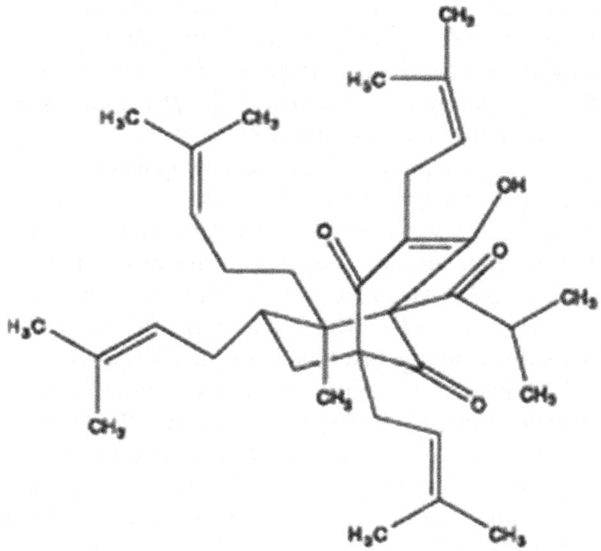

Fig. 4.3 Chemical structure of hyperforin

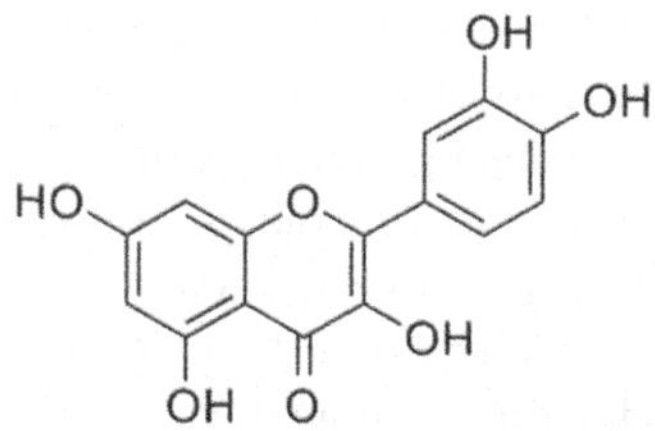

Fig. 4.4 Chemical structure of quercetin

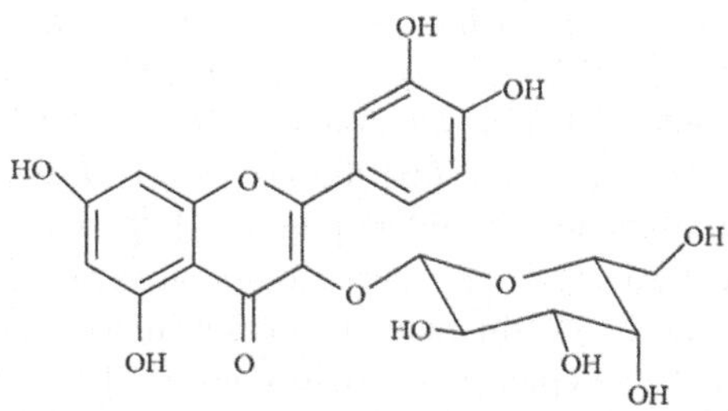

Fig. 4.5 Chemical structure of hyperoside

4.3 Phenolic Compound Content Studies on the *Hypericum* Species in Turkey

Phenolic compound contents of *H.androsaemum* L., *H.aviculariifolium* Jaub. and Spach, *H.aviculariifolium* Jaup. and Spach subsp., *depilatum, H. bithynicum* Boiss., *H. bupleuroides, H. calycinum* L., *H. heterophyllum* Vent., *H. hirsutum* L., *H. hyssopifolium* Vill., *H. hyssopifolium* L., *H. linarioides Bosse, H. lydium*

Boiss., H. montanum L., *H. montbretii* Spach, *H.nummularioides* Trautv., *H. orientale* L., *H. origanifolium* Willd., *H. perfoliatum* L., *H. perforatum* L., *H. pruinatum* Boiss., and Balansa, *H. scabrum* L., *H. triquetrifolium* Turra, *H. venustum* and *H. Xylosteifolium, Hypericum* species from Northern Turkey have been determined (Table 4.3).

Hypericin and pseudohypernicin compounds have been detected in all the species examined. Other compounds such as hyperforin, adhyperforin, hyperoside, chlorogenic acid, neochlorogenic acid, caffeic acid, 2,4 dihydroxybenzoic acid, isoquercitrin, quercitrin, quercetin, avicularin, rutin, 13,118-biapigenin, (+)-catechin, apig-7-*O*-glucoside, caffeic acid, amentoflavone, viteksin, and isoquercetin have also been identified (Table 4.3). The flavonoid vitexin was determined in *H. motbretii* for the first time in the genus *Hypericum* (Çirak et al. 2007d).

The hyperforin compound was found in *H. androsaemum, H. aviculariifolium, H. bithynicum, H. heterophyllum, H. montbretii, H. nummularioides, Hypericum orientale, H. origanifolium, H. perfoliatum, H. pruinatum,* and *H. scabrum* (Smelcerovic et al. 2008), and hypericin and pseudohypericin were detected in *H. orientale and H. scabrum* (Ayan and Çirak 2008). The presence of hypericin *in H.aviculariifolium subsp. deilatum, H. nummularoides, H. pruinatum, and H. venustum* was reported by Ayan et al. (2004) for the first time.

Smelcerovic et al. (2008) collected 17 wild-growing species of *Hypericum* plants (flowers) from various locations in Northern Turkey. They determined all extracts contained pseudohypericin, hypericin, hyperforin, rutin, hyperoside, quercitrin, and quercetin compounds. But rutin was not found in the extracts of *H. origanifolium* and *H. scabrum* plants. They reported that the greatest content of hypericin and pseudohypericin was observed in *H. perforatum* and *H. triquetrifolium* extracts. Hyperforin content was highest in the H. perforatum extract. Hyperforin was found for the first time in *H. androsaemum, H. aviculariifolium, H. bithynicum, H. heterophyllum, H. montbretii, H. nummularioides, Hypericum orientale, H. origanifolium, H. perfoliatum, H. pruinatum, and H. scabrum.* Thus, the greatest content of rutin was found in *H. montbretii* (Smelcerovic et al. 2008).

It was found that the *H. polyphyllum* Boiss. and Bal. contain several phenolic compounds pseudohypericin, hypericin, hyperforin, adhyperforin, hyperoside, chlorogenic acid, neochlorogenic acid, caffeic acid, rutin, 2,4-dihydroxybenzoic acid, 13,118-biapigenin, quercitrin, isoquercitrin, quercetin, avicularin, (−)-catechin and (−)-epicatechin (Cirak et al. 2017). Hyperforin, hypericin, pseudohypericin, chlorogenic acid, rutin, hyperoside, apigenin-7-*O*-glucoside, kaempferol, quercitrin, quercetin, and amentoflavone were detected in *H. scabrum* L. and *H. bupleuroides* Gris. It was reported that the phloroglucinol derivative hyperforin was not accumulated in *H. scabrum* (Ayan et al. 2009) (Table 4.4).

In research, *H. hircinum* L. subsp. Majus (Ainton) did not produce quercitrin, avicularin, or rutin and *H. lanuginosum* did not accumulate hyperforins, rutin, and caffeic acid in investigated hyperforin, adhyperforin, hypericin, pseudohypericin, chlorogenic acid, neochlorogenic acid, rutin, caffeic acid, hyperoside, 2,4 dihydroxybenzoic acid, isoquercitrin, quercetin, isoquercetin, avicularin, 13,118-biapigenin, (+)-catechin, (−)-epicatechin, compounds. Besides it was

Table 4.3 Studies on the phenolic compounds of *Hypericum* Species from Northern Turkey

Hypericum species	Examined phenolics	Method	References
H. androsaemum L.	Hypericin, pseudohypericin, hyperforin, adhyperforin, hyperoside, rutin, quercitrin, isoquercitrin, quercetin, chlorogenic acid, neochlorogenic acid, caffeic acid, 2,4 dihydroxybenzoic acid, avicularin, 13,118-biapigenin, (+)-catechin, (−)-epicatechin	LC/MS/MS, HPLC	Smelcerovic et al. (2008); Cirak et al. (2017)
H. aviculariifolium Jaub. and Spach	Hypericin, pseudohypericin, hyperforin, hyperoside, rutin, quercitrin, quercetin	LC/MS/MS	Smelcerovic et al. (2008)
H. montanum L.			
H. nummularioides Trautv.			
*H. aviculariifolium Jaup. and Spach subsp., depilatum**	Hypericin, pseudohypericin, chlorogenic acid, hyperoside, rutin, quercitrin, quercetin, apigenin-7-*O*-glucoside, hyperforin, isoquercetin	HPLC	Çirak et al. (2006, 2007d); Cirak et al. 2013)
H. bithynicum Boiss.	Hypericin, pseudohypericin, hyperforin, adhyperforin, chlorogenic acid, neochlorogenic acid, caffeic acid, 2,4 dihydroxybenzoic acid, hyperoside, isoquercitrin, quercitrin, quercetin, avicularin, rutin, 13,118-biapigenin, (+)-catechin	HPLC, LC/MS/MS	Cirak et al. (2016a); Smelcerovic et al. (2008)
*H. bupleuroides**	Hypericin, pseudohypericin, chlorogenicacid, hyperoside, rutin, hyperforin, quercetin, kaempferol, quercitrin, amentoflavon, apigenin-7-*O*-glucoside	HPLC	Ayan et al. (2009)

(continued)

Table 4.3 (continued)

Hypericum species	Examined phenolics	Method	References
H. calycinum L.	Hypericin, pseudohypericin, hyperforin, adhyperforin, chlorogenic acid, neochlorogenic acid, caffeic acid, 2,4 dihydroxybenzoic acid, hyperoside, isoquercitrin, quercitrin, quercetin, avicularin, rutin, 13,118-biapigenin, (+)-catechin	HPLC	Cirak et al. (2016a)
H. heterophyllum Vent.	Hypericin, pseudohypericin, hyperforin, hyperoside, rutin, quercitrin, quercetin	LC/MS/MS, HPLC	Smelcerovic et al. (2008); Ayan and Çirak (2008)
H. hirsutum L.	Hypericin, pseudohypericin, hyperforin, hyperoside, rutin, quercitrin, quercetin, adhyperforin, chlorogenic acid, neochlorogenic acid, caffeic acid, 2,4 dihydroxybenzoic acid, isoquercitrin, avicularin, rutin, 13,118-biapigenin, (+)-catechin	LC/MS/MS, HPLC	Smelcerovic et al. (2008); Cirak et al. (2016a)
H. hyssopifolium Vill.	Hypericin, pseudohypericin, hyperforin, hyperoside, rutin, quercitrin, quercetin,	LC/MS/MS, HPLC	Smelcerovic et al. (2008); Ayan et al. (2004)
H. hyssopifolium L.	Hypericin, pseudohypericin	HPLC	Ayan and Çirak (2008)
H. linarioides Bosse	Hypericin, pseudohypericin, hyperforin, hyperoside, rutin, quercitrin, quercetin,	HPLC, LC/MS/MS	Ayan and Çirak (2008) ; Smelcerovic et al. (2008)
H. lydium Boiss.	Hypericin, pseudohypericin, chlorogenic acid, neochlorogenic acid,	HPLC, UV spectrophotometer	Çirak (2006); Çirak et al. (2007d); Cirak et al. (2015)

(continued)

Table 4.3 (continued)

Hypericum species	Examined phenolics	Method	References
	hyperforin, adhyperforin, hyperoside, rutin, quercitrin, quercetin, apigenin-7-*O*-glucoside, 2,4 dihydroxybenzoic acid, isoquercitrin, avicularin, (+)-catechin, (−)-epicatechin, avicularin		
H. montbretii Spach	Hypericin, pseudohypericin, chlorogenic acid, hyperforin, hyperoside, rutin, quercitrin, isoquercetin quercetin, viteksin, apigenin-7-*O*-glucoside	HPLC, LC/MS/MS	Odabas et al. (2008); Ayan and Çirak (2008); Smelcerovic et al. (2008); Cirak and Radušienė (2007); Çirak et al. (2007d)
H. orientale L.	Hypericin, pseudohypericin, hyperforin, adhyperforin, chlorogenic acid, hyperoside, rutin, quercitrin, quercetin, apigenin-7-*O*-glucoside, isoquercetin, kaempferol, neochlorogenic acid, caffeic acid, 2,4 dihydroxybenzoic acid, avicularin, 13,118-biapigenin, (−)-epicatechin, rutin, amentoflavone	HPLC, LC/MS/MS	Çirak et al. (2007d, 2008, 2012, 2013); Smelcerovic et al. (2008); Camas et al. (2014a)
H. origanifolium Willd.	Hypericin, pseudohypericin, chlorogenic acid, hyperforin, hyperoside, quercitrin, isoquercetin, isoquercitrin, quercetin, rutin, apigenin-7-*O*-glucoside, caffeic acid	HPLC, LC/MS/MS	Ayan and Çirak (2008); Odabas et al. (2008); Smelcerovic et al. (2008); Çirak et al. (2007b, d)
H. perfoliatum L.	Hypericin, pseudohypericin, chlorogenic acid, hyperforin, hyperoside, quercitrin, isoquercetin, isoquercitrin, quercetin,	HPLC, LC/MS/MS	Odabas et al. (2008); Smelcerovic et al. (2008); Çirak et al. (2007a, d)

(continued)

Table 4.3 (continued)

Hypericum species	Examined phenolics	Method	References
	rutin, caffeic acid, apigenin-7-*O*-glucoside		
H. perforatum L.	Hypericin, pseudohypericin, chlorogenic acid, hyperforin, hyperoside, rutin, quercitrin, quercetin, apigenin-7-*O*-glucoside	LC/MS/MS, HPLC	Smelcerovic et al. (2008); Çirak et al. (2006, 2007c, d, 2008); Ayan and Çirak (2008)
H.pruinatum Boiss. and Balansa	Hypericin, pseudohypericin, chlorogenic acid, hyperforin, hyperoside, rutin, quercitrin, quercetin, apigenin-7-*O*-glucoside	HPLC, LC/MS/MS	Çirak et al. (2006, 2007a); Smelcerovic et al. (2008)
H. scabrum L.	Hypericin, pseudohypericin, chlorogenic acid, hyperoside, rutin, hyperforin, quercetin, kaempferol, quercitrin, amentoflavon, apigenin-7-*O*-glucoside	HPLC, LC/MS/MS	Ayan and Çirak (2008); Ayan et al. (2009); Smelcerovic et al. (2008)
H. triquetrifolium Turra	Hypericin, pseudohypericin, hyperforin, chlorogenic acid, hyperoside, rutin, quercitrin, quercetin, rutin, amentoflavone, kaempferol, apig-7-*O*-glucoside	LC/MS/MS, HPLC	Smelcerovic et al. (2008); Ayan and Çirak (2008); Cirak et al. (2014)
H. venustum	Hypericin, pseudohypericin, hyperforin	HPLC	Ayan et al. (2004)
H. xylosteifolium	Hypericin, pseudohypericin, hyperforin, adhyperforin, chlorogenic acid, neochlorogenic acid, caffeic acid, 2,4 dihydroxybenzoic acid, hyperoside, isoquercitrin, quercitrin, quercetin, avicularin, rutin, 13,118-biapigenin, (+)-catechin	HPLC	Cirak et al. (2016a)

found that the *H. pallens* Banks et Sol., have all of these compounds (Odabas et al. 2016).

Hypericin and pseudohypericin, chlorogenic acid, neochlorogenic acid, hyperforin, adhyperforin, caffeic acid, rutin, 2,4-dihydroxybenzoic acid, hyperoside, quercitrin, isoquercitrin, quercetin, avicularin, (−) –catechin, (−)-epicatechin, and amentoflavone compounds were investigated from *H. scabrum* L., *H. thymifolium* Banks and Sol., *H. lydium* Boiss., *H. olivieri* (Spach) Boiss., *H. linarioides* Bosse, *H. confertum* Choisy. It was found that the *H. thymifolium, H. linarioides, H. olivieri* and *H. scabrum* did not accumulate hyperforin and adhyperforin and those compounds were detectable only in flowers of *H. confertum* and *H. lydium* (Camas et al. 2014b).

Cirak et al. (2016b) investigated eight *Hypericum* species; *H. capitatum var. capitatum, H. capitatum* var. luteum, *H. elongatum* var. *elongatum, H. olympicum, H. polyphyllum, H. retusum, H. salsolifolium,* and *H. spectabile*, which are native to Southern Turkey. The compounds hyperforin, adhyperforin, hypericin, pseudohypericin, hyperoside, chlorogenic acid, neochlorogenic acid, caffeic acid, 2,4-dihydroxybenzoic acid, quercitrin, isoquercitrin, quercetin, avicularin, rutin, (+)-catechin, (−)-epicatechin, mangiferin, 13,118-biapigenin, and amentoflavone were determined from these species for the first time. They reported that the caffeic acid was not accumulated in *H. capitatum var. capitatum, H. capitatum var. luteum, H. retusum, H. spectabile, H. elongatum var. elongatum,* and *H. salsolifolium, similarly H. olympicum, H. polyphyllum*, did not produce hyperforin, adhyperforin and mangiferin. It was shown that the *H. elongatum* var. *elongatum* did not produce (+)-catechin (Cirak et al. 2016b).

The detailed chemical profiles of *H. aviculariifolium* subsp. *aviculariifolium* var. *albiflorum, H. cardiophyllum* and *H. elongatum subsp. microcalycinum* were made for the first time. Hyperforin, adhyperforin, hypericin, pseudohypericin, hyperoside, chlorogenic acid, neochlorogenic acid, caffeic acid, 2,4 dihydroxybenzoic acid, quercitrin, isoquercitrin, quercetin, avicularin, rutin, 13,118-biapigenin, and (+)-catechin contents were investigated of these species. It was reported that *H. cardiophyllum* accumulated the highest hyperoside and quercetin levels (Cirak et al. 2016a) (Table 4.5).

The researches from other regions of Turkey especially on endemic species showed that *H. neurocalycinum* was rich with chlorogenic acid, rutin, kaempferol, quercetin, and amentoflavon. *H. malatyanum* was rich with hyperoside, pseudohypericin, quercitrin, and isoquercitrin (Ozkan et al. 2018). Hyperoside and isoquercetin were determined as the main compounds of the *H. capitatum* Choisy var. *Capitatum*, but rutin and amentoflavon were not detected in these species (Boga et al. 2016). Hypericin, pseudohypericin, hyperoside, chlorogenic acid, rutin, quercitrin, quercetin, kaempferol, apigenin-7-*O*-glucoside, and amentoflavone chemicals were detected at various levels in *H. adenotrichum* (Çirak et al. 2009). It was determined that the aerial plant parts of *H. leptophyllum* contain hyperoside, chlorogenic acid, quercitrine, isoquercetine, and quercetine, but they did not have hypericin, hyperforin, pseudohypericin, rutin, and kaempferol.

Table 4.4 Studies on the phenolic compounds of *Hypericum* Species from Southern Turkey

Hypericum species	Examined phenolics	Method	References
H. aviculariifolium subsp. *aviculariifolium* var. *albiflorum*	Hyperforin, adhyperforin, hypericin, pseudohypericin, hyperoside, chlorogenic acid, neochlorogenic acid, caffeic acid, rutin, 2,4 dihydroxybenzoic acid, quercitrin, isoquercitrin, quercetin, avicularin, 13,118-biapigenin, (+)-catechin	HPLC	Cirak et al. (2016a)
H. elongatum subsp. *microcalycinum*			
H. cardiophyllum			
H. capitatum var. *capitatum*	Hyperforin, adhyperforin, hypericin, pseudohypericin, hyperoside, chlorogenic acid, neochlorogenic acid, rutin, caffeic acid, 2,4 dihydroxybenzoic acid, isoquercitrin, quercitrin, quercetin, avicularin, (+)-catechin, 13,118-biapigenin, amentoflavone, mangiferin	HPLC	Cirak et al. (2016b)
H. capitatum var. *luteum*			
H. elongatum var. *elongatum*			
H. spectabile			
H. salsolifolium			
H. retusum			
H. olympicum			
H. hircinum L. subsp. Majus	Hyperforin, adhyperforin, hypericin, pseudohypericin, hyperoside, chlorogenic acid, neochlorogenic acid, rutin, caffeic acid, 2,4 dihydroxybenzoic acid, isoquercitrin, quercetin, isoquercetin, avicularin, 13,118-biapigenin, (+)-catechin, ()-epicatechin	*HPLC*	Odabas et al. (2016)
H. lanuginosum Lam.			
H. russeggeri (Fenzl)			
H. confertum Choisy	Hyperforin, adhyperforin, hypericin, pseudohypericin, hyperoside, chlorogenic acid, neochlorogenic acid, rutin, caffeic acid, 2,4 dihydroxybenzoic acid, isoquercitrin, quercitrin, quercetin, avicularin, (+)-catechin, amentoflavone	HPLC	Camas et al. (2014b)
H. thymifolium Banks and Sol.			
H. olivieri (Spach) Boiss.			
H. linarioides Bosse			
H. scabrum L.			
H. lydium Boiss.	Hyperforin, adhyperforin, hypericin, pseudohypericin, hyperoside, chlorogenic acid, neochlorogenic acid, rutin, caffeic acid, 2,4 dihydroxybenzoic acid, isoquercitrin, quercitrin, quercetin, avicularin, amentoflavone, (+)-catechin, (−)-epicatechin,	HPLC	Camas et al. (2014b); Cirak et al. (2015)
H. pallens Banks et Sol.	Hyperforin, adhyperforin, hypericin, pseudohypericin, hyperoside, chlorogenic acid, neochlorogenic acid, caffeic acid, rutin, 2,4 dihydroxybenzoic acid, isoquercitrin, quercitrin, quercetin,	*HPLC*	Odabas et al. (2016); Camas et al. (2014a)

(continued)

Table 4.4 (continued)

Hypericum species	Examined phenolics	Method	References
	avicularin, amentoflavone 13,118-biapigenin, (+)-catechin, (−)-epicatechin, isoquercetin		
H. polyphyllum	Hyperforin, adhyperforin, hypericin, pseudohypericin, hyperoside, chlorogenic acid, neochlorogenic acid, rutin, caffeic acid, 2,4 dihydroxybenzoic acid, isoquercitrin, quercitrin, quercetin, avicularin, 13,118-biapigenin, (+)-catechin, (−)-epicatechin, amentoflavone, mangiferin	HPLC	Cirak et al. (2016b, 2017)

Table 4.5 Studies on the phenolic compounds of *Hypericum* species from other regions of Turkey

Hypericum species	Region	Examined phenolics	Method	References
H. malatyanum	Eastern Anatolia	Hypericin, pseudohypericin, rutin chlorogenic acid, ısoquercitrin, quercetin, kaempferol, amentoflavon	HPLC	Ozkan et al. (2018)
H. neurocalycinum	Middle Anatolia			
H.capitatum Choisy var.*capitatum*	Western Anatolia	Hypericin, pseudohypericin, hyperforin, chlorogenic acid, hyperoside, quercetin ısoquercitrin, kaempferol, quercitrin	HPLC	Boga et al. (2016)
H. adenotrichum	Western Anatolia	Hypericin, pseudohypericin, hyperoside chlorogenic acid, rutin, quercitrin, quercetin, amentoflavone, kaempferol, apigenin-7-*O*-glucoside	HPLC	Çirak et al. (2009)
H. leptophyllum	Middle Anatolia	Chlorogenic acid, hyperoside, quercitrin, quercetin	HPLC	Camas et al. (2012)

4.4 Biological Activity of *Hypericum* Species from Turkey

Historically plants have always had important roles in the improvement of human health. People have relied on them for their therapeutic and preventive uses because they are natural substances. One of these plants is *H. perforatum*, which is of medical importance (Table 4.6).

It was revealed that some major constituents were responsible for the biological activity of *H. perforatum*. For example it was reported that anti-inflammatory, antiulcerogenic (Berghoefer and Hoelzl 1989) activity of amentoflavone;

Table 4.6 Recent studies on the biological activity of *Hypericum* species growing in Turkey

Hypericum species	Biological activities	References
H. aviculariifolium Jaub. and Spach subsp. *depilatum*	Antioxidant, antibacterial	Maltas et al. (2013)
H. capitatum var. *capitatum*	Antioxidant, antimicrobial, anticholinesterase, DNA damage inhibitor	Boga et al. (2016)
H. confertum subsp. confertum	Antimicrobial	Kunduhoglu et al. (2011)
H. heterophyllum	Cytotoxic properties, mitotic, proliferative	Öcal and Eroğlu (2012)
H. lydium	Antioxidant, antimutagenic	Boran and Ugur (2017)
H. montbretii Spach.	Immunomodulating inhibitory activity	Demirkiran et al. (2013)
	Sedative and anticonvulsant	Can and Özkay (2012)
	Antidepressant	Can et al. (2011a, b, c)
H. neurocalycinum	Antioxidant, antimicrobial	Özkan et al. (2013a)
H. pamphylicum	Antioxidant, antimicrobial	Özkan et al. (2013b)
	Memory-vitalizing	Senol et al. (2016)
H. perforatum	Antioxidant	Altun et al. (2013); Maltas et al. (2013); Eğilmez et al. (2015); Uslusoy et al. (2017a, b); Nazıroğlu et al. (2014)
	Antibacterial	Maltas et al. (2013)
	Wound healing	Kiyan et al. (2015); Dikmen et al. (2011); Han et al. (2017); Damlar et al. (2017)
	Antipsychotic	Gurok et al. (2014)
	Depression treatment	Can et al. (2011a, b, c); Dalmizrak et al. (2012)
	Antitumoral	Ocak et al. (2013)
	Antidiabetic	Can et al. (2011a, b, c)
	Antiinflammatory	Nazıroğlu et al. (2014)
	Antiapoptotic	Ocak et al. (2013)
H. salsugineum Robson and Hub.-Mor.,	Antioxidant, antibacterial	Maltas et al. (2013)
H. uniglandulosum	Antioxidant, antibacterial	Turkoglu et al. (2015)

antibacterial and wound-healing (Gurevich et al. 1971; Bystrov et al. 1975; Maisenbacher and Kovar 1992), neurotransmitter inhibitor (Nahrstedt and Butterweck 1997), antidepressant (Chatterjee et al. 1998), potential anticarcinogenic (Schempp et al. 2002), angiogenesis inhibition (Schempp et al. 2005) and antimalarial activity of hyperforin (Verotta et al. 2007); antiviral (Meruelo et al. 1988; Weber et al. 1994; Lavie et al. 1995), antidepressant activity (Nahrstedt and Butterweck 1997; Chatterjee et al. 1998; Müller et al. 2001; Eckert and Müller 2001), anti-inflammatory activity of hypericin (Panossian et al. 1996); sedative effect of 13,118-biapigenin (Berghoefer and Hoelzl 1989); antiviral activity of pseudohypericin (Meruelo et al. 1988); in vitro MAO-inhibiting activity of quercitrin (Sparenberg et al. 1993); antidepressant, antimicrobial, antiviral, diuretic, cardiotonic, MAOA inhibitor activities of xanthones (Kitanov and Blinova 1987; Hölzl et al. 1989; Sparenberg et al. 1993; Nahrstedt and Butterweck 1997).

Numerous studies were conducted for determining the biological activities of the *Hypericum* species growing in Turkey. It was found *H. perforatum* L. has acetylcholinesterase and butyrylcholinesterase inhibition activity that related to Alzheimer's and Parkinson's disease (Altun et al. 2013). Some studies on the rats show that *H. perforatum* has hepatoprotective (Bayramoglu et al. 2014), protective effect on SCI-induced oxidative stress and apoptosis (Özdemir et al. 2016) and increased the levels of Zn and Na in serum, and Cd levels in the hair (Ekin et al. 2012). Additionally, some clinical studies revealed that *H. perforatum* shows the preventive effects on intestinal ischemia-reperfusion injury in hamsters (Ocak et al. 2014), on oxidative stress and [Ca2+] ion concentrations by modulating transient receptor potential and voltage-gated calcium channel in the patients with MS (Naziroglu et al. 2014), on DNA damage, cell viability, apoptosis and a disintegrin-like and metalloproteinase (Akyol et al. 2014). Studies revealed that *H. lydium* may be provided as a useful and safe agent for the prevention of cancer and mutations (Boran and Ugur 2017).

Hypericum species have active antibacterial components such as hyperforin and flavonoids, which probably contribute to rapid and better wound healing (Shetty et al. 2008). It is well known that flavonoids have an active antioxidant capacity for the inhibition of lipid peroxidation, which leads to the prevention of cell damage and an increase in the viability of collagen fibrils (Kiyan et al. 2015).

Can et al. (2011a, b, c) showed that *H. montbretti* has antidepressant activities that reported from originating from rutin. Additionally, the anti-inflammatory and anti-nociceptive activities of *H. triquetrifolium* was known (Apaydın et al. 1999; Ozturk et al. 2002). *H. calycinum* may be used for therapeutic purposes in depression since it has the anti-depressant effect that was as potent as that of *H. perforatum* (Smelcerovic et al. 2008). The *hypericum* species, *H. lysimachioides* var. *lysimachioides*, *H. hyssopifolium* subsp. *elongatum* var. *elongatum*, *H. montbretii*, *H. venustum*, *H. pamphylicum*, and *H. perforatum* were found to have antioxidant properties and were investigated for antibacterial activity (Cakir et al. 2003; Eroglu 2007).

4.5 Conclusion

Our country is a significant gene center for *Hypericum species*. These species have been used for a long time in folk medicine. Several species among the 100 *Hypericum* taxa growing widely in Turkey were investigated, but there are many types of *Hypericum* species that have not yet been studied for their chemical contents. Recently, the studies focused on the genus *Hypericum*, because it is a source of a variety of compounds with different biological activities. Despite the quantitative changes observed in the chemical compound content of plants, it was found the phytochemical contents of the species from the same section were quite similar. In our country, although several *Hypericum* species have been used in folk medicine, only *H. perforatum* exists as its pharmaceuticals in the market.

The present study could be helpful to reveal new sources of phytochemicals for potential pharmaceutical studies on the *Hypericum species* that are growing in Turkey.

References

Akan H, Aydoğdu M, Korkut MM, Balos MM (2013) An ethnobotanical research of the Kalecik mountain area (Şanlıurfa, South-East Anatolia). Biodivers Conserv 6(2):84–90

Akgul A (2008) Ethnobotany at Midyat (Mardin). Master Thesis. Ege University, Institute of Science, Izmir

Aktan T (2011) Ethnobotanical studies of Yenisehir (Bursa) Villages. Master Thesis. Manisa, Celal Bayar University, Institute of Science, Department of Botany

Akyol S, Yükselten Y, Çakmak Ö, Uğurcu V, Altuntaş A, Gürler M, Demircan K (2014) Hydrogen peroxide-induced oxidative damage in human chondrocytes: the prophylactic effects of *Hypericum perforatum* Linn extract on deoxyribonucleic acid damage, apoptosis and matrix remodeling by a disintegrin-like and metalloproteinase with thrombospondin motifs proteinases. Arch Rheumatol 29:203–214

Apaydın S, Zeybek U, Ince I, Elgin G, Karamenderes C, Ozturk B, Tuglular I (1999) Hypericum triquetrifolium Turra. extract exhibits antinociceptive activity in the mouse. J Ethnopharmacol 67(3):307–312

Altun ML, Yılmaz BS, Orhan IE, Citoglu GS (2013) Assessment of cholinesterase and tyrosinase inhibitory and antioxidant effects of *Hypericum perforatum* L. (St. John's wort). Ind Crop Prod 43:87–92

Altundag E, Ozturk M (2011) Ethnomedicinal studies on the plant resources of East Anatolia, Turkey. Procedia Soc Behav Sci 19:756–777

Ayan AK, Çirak C (2008) Hypericin and Pseudohypericin contents in some *Hypericum* species growing in Turkey. Pharm Biol 46:288–291

Ayan AK, Cirak C, Kevseroglu K, Zen T (2004) Hypericin in some *Hypericum* species from Turkey. Asian J Plant Sci 3:202–204

Ayan AK, Radušienė J, Çirak C, Ivanauskas L, Janulis V (2009) Secondary metabolites of *Hypericum scabrum* and *Hypericum bupleuroides*. Pharm Biol 47:847–853

Bayramoglu G, Bayramoglu A, Engur S, Senturk H, Ozturk N, Colak S (2014) The hepatoprotective effects of *Hypericum perforatum* L. on hepatic ischemia/reperfusion injury in rats. Cytotechnology 66:443–448

Baytop T (1999) Therapy with medicinal plants in Turkey. Istanbul University Press, İstanbul, pp 66–167

Berghoefer R, Hoelzl J (1987) Biflavonoids in *Hypericum perforatum:* part 1. Isolation of I3, II8-biapigenin. Planta Med 53:216–217

Berghoefer R, Hoelzl J (1989) Isolation of I3, II8-biapigenin (amentoflavone) from *Hypericum perforatum*. Planta Med 55:91

Bilia AR, Bergonzi MC, Mazzi G, Vincieri FF (2001) Analysis of plant complex matrices by use of nuclear magnetic resonance spectroscopy: St. John's Wort extract. J Agric Food Chem 49:2115–2124

Bingol U, Cosge B, Gurbuz B (2011) Odabas MS, Cirak C (Eds.), *Hypericum* species in flora of Turkey. Med Aromat Plant Sci Biotechnol 5(special issue 1):86–90

Boga M, Ertas A, Eroglu-Ozkan E, Kizil M, Ceken B, Topcu G (2016) Phytochemical analysis, antioxidant, antimicrobial, anticholinesterase and DNA protective effects of *Hypericum capitatum* var. *capitatum* extracts. S Afr J Bot 104:249–257

Bombardelli E, Morazzoni P (1995) Hypericum perforatum. Fitoterapia 66:43–68

Boran R, Ugur A (2017) The mutagenic, antimutagenic and antioxidant properties of *Hypericum lydium*. Pharm Biol 55:402–405

Bruneton J (1995) Pharmacognosy, phytochemistry, medicinal plants. Lavoisier Publishing, Paris, pp 367–370

Bulut Y (2006) Useful plants of Manavgat District (Antalya). Master Thesis. Suleyman Demirel University, Institute of Science, Department of Biology, Isparta

Buyukgebiz T (2006) Non-wood Forest products of Sutculer District (Isparta), Master Thesis. Suleyman Demirel University, Department of Forest Engineering, Isparta

Bystrov NS, Chernov BK, Dobrynin VN, Kolosov MN (1975) The structure of hyperforin. Tetrahedron Lett 16:2791–2794

Cakir A, Mavi A, Yildirim A, Duru ME, Harmandar M, Kazaz C (2003) Isolation and characterization of antioxidant phenolic compounds from the aerial parts of *H. hyssopifolium* L. by activity-guided fractionation. J Ethnopharmacol 87:73–83

Camas N, Radusiene J, Stanius Z, Caliskan O, Cirak C (2012) Secondary metabolites of *Hypericum leptophyllum* Hochst., an endemic Turkish species. The Scientific World Journal, Singapore

Camas N, Radusiene J, Ivanauskas L, Jakstas V, Cirak C (2014a) Altitudinal changes in the content of bioactive substances in *Hypericum orientale* and *Hypericum pallens*. Acta Physiol Plant 36:675–686

Camas N, Radusiene J, Ivanauskas L, Jakstas V, Kayıkçı S, Cirak C (2014b) Chemical composition of *Hypericum* species from the taeniocarpium and drosanthe sections. Plant Syst Evol 300:953–960

Can ÖD, Özkay ÜD (2012) Effects of *Hypericum montbretti* extract on the central nervous system and involvement of GABA (A)/benzodiazepine receptors in its pharmacological activity. Phytother Res 26:1695–1700

Can ÖD, Ismail IB, Öztürk Y, Öztürk N, Potoğlu-Erkara İ, Sagratini G, Maggi F (2011a) New antidepressant drug candidate: *Hypericum montbretti* extract. Nat Prod Res 25:1469–1472

Can ÖD, Öztürk Y, Özkay ÜD (2011b) Effects of insulin and St. John's wort treatments on anxiety, locomotory activity, depression, and active learning parameters of streptozotocin-diabetic rats. Planta Med 77:1970–1976

Can ÖD, Öztürk Y, Öztürk N, Sagratini G, Ricciutelli M, Vittori S, Maggi F (2011c) Effects of treatment with St. John's Wort on blood glucose levels and pain perceptions of streptozotocin-diabetic rats. Fitoterapia 82:576–584

Cellarova E, Daxnerova Z, Kimakova K, Haluskova J (1994) The variability of the hypericin content in the regenerants of *Hypericum perforatum*. Acta Biotechnol 14:267–274

Chatterjee SS, Nöldner M, Koch E, Erdelmeier C (1998) Antidepressant activity of Hypericum perforatum and hyperforin: the neglected possibility. Pharmacopsychiatry 31(S 1):7–15

Cimen OD (2007) Ethnobotanical studies about traditional medicines in Konya. Master thesis, Gazi University, Institute of Health Science, program of Phytotheraphy. Ankara Cirak C. Hypericin in *H. lydium* growing in Turkey. Biochem Syst Ecol 34:897–899

Çirak C (2006) Hypericin in *Hypericum lydium* Boiss. Growing in Turkey. Biochem Syst Ecol 34:897–899

Cirak C, Radušienė J (2007) Variation of hyperforin in *Hypericum montbretii* during its phenological cycle. Nat Prod Res 21:1151–1156

Çirak C, Sağlam B, Ayan AK, Kevseroğlu K (2006) Morphogenetic and diurnal variation of hypericin in some *Hypericum* species from Turkey during the course of ontogenesis. Biochem Syst Ecol 34:1–13

Çirak C, Radušienė J, Janulis V, Ivanauskas L (2007a) Secondary metabolites in *Hypericum perfoliatum:* variation among plant parts and phenological stages. Bot Helv 117:29–36

Çirak C, Radušienė J, Ivanauskas L, Janulis V (2007b) Variation of bioactive secondary metabolites in *Hypericum origanifolium* during its phenological cycle. Acta Physiol Plant 29:197–203

Çirak C, Radušienė J, Karabük BS, Janulis V (2007c) Variation of bioactive substances and morphological traits in *Hypericum perforatum* populations from northern Turkey. Biochem Syst Ecol 35:403–409

Çirak C, Radusiene J, Janulis V, Ivanauskas L, Arslan B (2007d) Chemical constituents of some *Hypericum* species growing in Turkey. J Plant Biol 50:632–635

Çirak C, Radusiene J, Janulis V, Ivanauskas L (2008) Pseudohypericin and hyperforin in *Hypericum perforatum* from northern Turkey: variation among populations, plant parts and phenological stages. J Integr Plant Biol 50:575–580

Çirak C, Ivanauskas L, Janulis V, Radušienė J (2009) Chemical constituents of *Hypericum adenotrichum* Apach, an endemic Turkish species. Natural Prod Res Part A: Struct Synth 23:1189–1195

Cirak C, Radusiene J, Stanius Z, Camas N, Caliskan O, Odabas MS (2012) Secondary metabolites of *Hypericum orientale* L. growing in Turkey: variation among populations and plant parts. Acta Physiol Plant 34:1313–1320

Cirak C, Radusiene J, Camas N, Caliskan O, Odabas MS (2013) Changes in the contents of main secondary metabolites in two Turkish *Hypericum* species during plant development. Pharm Biol 51:391–399

Cirak C, Radusiene J, Ivanauskas L, Jakstas V, Çamaş N (2014) Changes in the content of bioactive substances among *Hypericum montbretii* populations from Turkey. Rev Bras 24:20–24

Cirak C, Radusiene J, Ivanauskas L, Jakstas V, Çamaş N (2015) Population variability of main secondary metabolites in *Hypericum lydium* Boiss. (Hypericaceae). Iran J Pharmaceut Res 14:969

Cirak C, Radusiene J, Jakstas V, Ivanauskas L, Seyis F, Yayla F (2016a) Secondary metabolites of seven *Hypericum* species growing in Turkey. Pharm Biol 54:1–10

Cirak C, Radusiene J, Jakstas V, Ivanauskas L, Yayla F, Seyis F, Camas N (2016b) Secondary metabolites of *Hypericum* species from the Drosanthe and Olympia sections. S Afr J Bot 104:82–90

Cirak C, Radusiene J, Jakstas V, Ivanauskas L, Seyis F, Yayla F (2017) Altitudinal changes in secondary metabolite contents of *Hypericum androsaemum* and *Hypericum polyphyllum*. Biochem Syst Ecol 70:108–115

Crockett SL, Robson NKB (2011) Odabas, M.S., Cirak, C. (Eds.), Taxonomy and chemotaxonomy of the genus *Hypericum*. Medicin Aromat Plant Sci Biotechnol 5(special issue 1):1–13

Dalmizrak O, Kulaksiz-Erkmen G, Ozer N (2012) Evaluation of the in vitro inhibitory impact of Hypericin on placental glutathione S-transferase pi. -the. Protein J 31:544–549

Damlar I, Arpağ OF, Tatli U, Altan A (2017) Effects of *Hypericum perforatum* on the healing of xenografts: a histomorphometric study in rabbits. Br J Oral Maxillofac Surg 55:383–387

Davis PH (1988) Flora of Turkey and the East Aegean Islands. Edinburgh University Press, Edinburgh

Davis PH, Cullen J, Coode MJE (1967) Flora of Turkey and the East Aegean Islands, vol 2. Edinburgh University Press, Edinburgh, p 581

Demirci S (2010) Ethnobotanical study in Andirin (Kahramanmaras) district. Master Thesis. Istanbul University, Institute of Health Science, Department of Pharmaceutical Botany, Istanbul

Demirkiran O, Mesaik MA, Beynek H, Abbaskhan A, Choudhary MI (2013) Immunosupressive phenolic constituents from *Hypericum montbretii* Spach. Record Nat Prod 7:210

Dikmen M, Öztürk Y, Sagratini G, Ricciutelli M, Vittori S, Maggi F (2011) Evaluation of the wound healing potentials of two subspecies of *Hypericum perforatum* on cultured NIH3T3 fibroblasts. Phytother Res 25:208–214

Ecevit GG, Ozhatay N (2006) An ethnobotanical study in Catalca (European part of Istanbul) II. Turkish J Pharmaceut Sci 3:73–89

Eckert GP, Müller WE (2001) Effects of hyperforin on the fluidity of brain membranes. Pharmacopsychiatry 34:(Sup. 1)22–(Sup. 1)25

Eğilmez OK, Kökten N, Ekici AID, Kalcıoğlu MT, Yesilada E, Tekin M (2015) The effect of *Hypericum perforatum* L.(St. John's wort) on prevention of myringosclerosis after myringotomy in a rat model. Int J Pediatr Otorhinolaryngol 79:1128–1134

Ekim T, Koyuncu M, Vural M, Duman H, Aytaç Z, Adıgüzel N (2000) Red data book of Turkish plants (Pteridophyta and Spermatophyta), Türkiye Tabiatını Koruma Dernegi Van 100. Yıl Üniversitesi, Barıçcan Ofset, Ankara, p 246

Ekin S, Oto G, Yardim Y, Ozgokce F, Kusman T (2012) Protective effect of *Hypericum perforatum* L. on serum and hair trace elements in rats 7, 12-dimethylbenz [a] anthracene-induced oxidative stress. Environ Toxicol Pharmacol 33:440–445

Eroglu E (2007) Investigaton of the H. pamphylicum in the sense of hypericin content and biological activity. Master Thesis, Istanbul University, Department of Pharmacognosy

Ezer N, Avcı K (2004) Traditional medicines of cerkes (Cankiri) region. J Hacet Univ, Fac Pharm 24(2):67–80

Ezer N, Mumcu AO (2006) Folk Medicines in Merzifon (Amasya, Turkey). Turk J Bot 30:223–230

Fiebich BL, Knorle R, Apel K, Kammler T, Weiss G (2011) Pharmacological studies in an herbal drug combination of St. John's wort (*Hypericum perforatum*) and passion flower (*Passiflora incarnata*). In vitro and in vivo evidence of synergy between *Hypericum* and *Passiflora* in antidepressant pharmacological models. Fitoterapia 82:474–480

Güner A, Özhatay N, Ekim T, Baçer KHC (2000) Flora of Turkey and the East Aegean Islands (supplement), vol 11. Edinburgh University Press, Edinburgh, p 656

Guner A, Aslan S, Ekim T, Vural M, Babaç MT (2012) List of Turkish flora vascular plants. Available online at: publication of Nezahat Gokyigit botanical garden and flora research Fondation, Istanbul. http://www.bizimbitkiler.org.tr/v2/hiyerarsi.php. *Hypericum*

Gurevich AI, Dobrynin VN, Kolosov MN, Popravko SA, Riabova ID (1971) Antibiotic hyperforin from Hypericum perforatum L. Antibiotiki 16:510–513

Gurok MG, Mermi O, Kilic F, Canan F, Kuloglu M (2014) Psychotic episode induced by St. John's wort (*Hypericum Perforatum*): a case report. J Mood Disorders 4:38–40

Han MC, Durmuş AS, Sağlıyan A, Günay C, Özkaraca M, Kandemir FM, Öztopalan DF (2017) Effects of Nigella sativa and Hypericum perforatum on wound healing. Turk J Vet Anim Sci 41:99–105

Hansen SH, Jensen AG, Cornett C, Bjornsdottir I, Wright B, Wilson ID (1999) High-performance liquid chromatography on-line coupled to high-field NMR and mass spectrometry for structure elucidation of constituents of *Hypericum perforatum* L. Anal Chem 71:5235–5241

Hegnauer R (1989) Chemotaxonomie der Pflanzen, vol VII. Birkhauser Verlag, Berlin

Hobbs C (1989) St. John's wort. Herbal Gram 18(19):24–33

Hölzl J, Demisch L, Gollnik B (1989) Investigations about antidepressiva and mood changing effects of Hypericum perforatum. Planta Med 55(07):643–643

Kasper S, Caraci F, Forti B, Drago F, Aguglia E (2010) Efficacy and tolerability of *Hypericum* extract for the treatment of mild to moderate depression. Eur Neuropsychopharmacol 20:74–765

Keskin M, Alpınar K (2002) Ethnobotanical studies about Kislak (Yayladağı-Hatay). J Syst Bot 9:91–100

Kitanov GM, Blinova KF (1987) Modern state of the chemical study of species of the genus Hypericum. Chem Nat Compd 23(2):151–166

Kiyan S, Uyanikgil Y, Altunci YA, Cavusoglu T, Cetin Uyanikgil EO, Karabey F (2015) Investigation of acute effects of *Hypericum perforatum* (St. John's wort-Kantaron) treatment in experimental thermal burns and comparison with silver sulfadiazine treatment. Turkish J Traum Emerg Surg 21:323–336

Kizilarslan C (2008) An ethnobotanical survey in the south part of İzmit gulf. Master Thesis. Istanbul University, Institute of Health Science, Department of Pharmaceutical Botany, Istanbul

Kultur S (2007) Medicinal plants used in Kirklareli Province (Turkey). J Ethnopharmacol 111:341–364

Kunduhoglu B, Pilatin S, Caliskan F (2011) Antimicrobial screening of some medicinal plants collected from Eskisehir, Turkey. Fresenius Environ Bull 20:945–952

Kurkin VA, Pravdivtseva OE (2007) Flavonoids from the aerial parts of *Hypericum perfortum*. Chem Nat Compd 43:620–621

Kurth H, Spreemann R (1998) Phytochemical characterization of various St. John's wort extracts. Adv Ther 15:117–128

Lavie G, Mazur Y, Lavie D, Meruelo D (1995) The chemical and biological properties of hypericin - a compound with a broad spectrum of biological activities. Med Res Rev 15:111–119

Maisenbacher P, Kovar KA (1992) Adhyperforin: a homologue of hyperforin from Hypericum perforatum. Planta Med 58(03):291–293

Maltas E, Uysal A, Yildiztugay E, Aladag MO, Yildiz S, Kucukoduk M (2013) Investigation of antioxidant and antibacterial activities of some *Hypericum* species. Fresenius Environ Bull 22:862–869

Mart S (2006) An ethnobotanical investigation of the natural plants using by inhabitants in Bahce and Hasanbeyli districts of Osmaniye Province. University of Cukurova, Institute of Natural and Applied Sciences, Department of Biology, Adana

Meruelo D, Lavie G, Lavie D (1988) Therapeutic agents with dramatic antiretroviral activity and little toxicity at effective doses: aromatic polycyclic diones hypericin and pseudohypericin. Proc Natl Acad Sci 85:5230–5234

Müller WE, Singer A, Wonnemman M (2001) Hyperforin: antidepressant activity by a novel mechanism of action. Pharmacopsychiatry 34:S98–S102

Naeem I, Saddiqe Z, Patel AV, Hellio C (2010) Flavonoid analysis and antibacterial activity of extracts of *Hypericum perforatum*. Asian J Chem 22:3596–3600

Nahrstedt A, Butterweck V (1997) Biologically active and other chemical constituents of the herb of *Hypericum perforatum* L. Pharmacopsychiatry 30:129–134

Nazıroğlu M, Şahin M, Çiğ B, Aykur M, Erturan İ, Ugan Y (2014) *Hypericum perforatum* modulates apoptosis and calcium mobilization through voltage-gated and TRPM2 calcium channels in neutrophil of patients with Behcet's disease. J Membr Biol 247:253–262

Naziroglu M, Kutluhan S, Övey İS, Aykur M, Yurekli VA (2014) Modulation of oxidative stress, apoptosis, and calcium entry in leukocytes of patients with multiple sclerosis by *Hypericum perforatum*. Nutr Neurosci 17:214–221

Ocak Z, Acar M, Gunduz E, Gunduz M, Demircan K, Uyeturk U, Ozlu T (2013) Effect of hypericin on the ADAMTS-9 and ADAMTS-8 gene expression in MCF7 breast cancer cells. Eur Rev Med Pharmacol Sci 17:1185–1190

Ocak T, Duran A, Özyalvaçli G, Ocak Z, Terzi EH, Tosun M, Erdem K (2014) Protective effects of montelukast and *Hypericum perforatum* against intestinal ischemia-reperfusion injury in hamsters. Turkish J Med Sci 44:381–386

Öcal A, Eroğlu HE (2012) In vitro cytogenetic effects of *Hypericum heterophyllum* in human peripheral blood lymphocytes. Bangladesh J Pharmacol 7:36–41

Odabas MS, Radusiene J, Cirak C, Camas N (2008) Prediction models for the phenolic contents in some *Hypericum* species from Turkey. Asian J Chem 20:4792

Odabas MS, Radusiene J, Ivanauskas L, Jakstas V, Camas N, Kayikci S (2016) Secondary metabolites in *Hypericum* species and their distribution in different plant parts. Žemdirbystė (Agriculture) 103:193–198

Özdemir ÜS, Nazıroğlu M, Şenol N, Ghazizadeh V (2016) *Hypericum perforatum* attenuates spinal cord injury-induced oxidative stress and apoptosis in the dorsal root ganglion of rats: involvement of TRPM2 and TRPV1 channels. Mol Neurobiol 53:3540–3551

Ozturk B, Apaydin S, Goldeli E, Ince I, Zeybek U (2002) Hypericum triquetrifolium Turra. extract exhibits antiinflammatory activity in the rat. J Ethnopharmacol 80(2–3):207–209

Özkan EE, Nurten O, Ozhan G, Celik BO, Mat A (2013a) Chemical compositions and biological activities of *Hypericum neurocalycinum* an endemic species of Turkey. Planta Med 79:PJ14

Özkan EE, Özsoy N, Özhan G, Celik BÖ, Mat A (2013b) Chemical composition and biological activities of *Hypericum pamphylicum*. Ind Crop Prod 50:182–189

Ozkan EE, Ozden TY, Ozsoy N, Mat A (2018) Evaluation of chemical composition, antioxidant and anti-acetylcholinesterase activities of *Hypericum neurocalycinum* and *Hypericum malatyanum*. S Afr J Bot 114:104–110

Panossian AG, Gabrielian E, Manvelian V, Jurcic K, Wagner H (1996) Immunosuppressive effects of hypericin on stimulated human leukocytes: inhibition of the arachidonic acid release, leukotriene B4 and Interleukin-Iα production, and activation of nitric oxide formation. Phytomedicine 3(1):19–28

Pietta P, Gardana C, Pietta A (2001) Comparative evaluation of St. John's wort from different Italian regions. Farmaco 56:491–496

Schempp CM, Kirkin V, Simon-Haarhaus B, Kersten A, Kiss J, Termeer CC (2002) Inhibition of tumour cell growth by hyperforin, a novel anticancer drug from St. John's wort that acts by induction of apoptosis. Oncogene 21:1242–1250

Schempp CM, Kiss J, Kirkin V (2005) Hyperforin acts as an angiogenesisinhibitor. Planta Med 71:999–1004

Schey KL, Patat S, Chignell CF, Datillo M, Wang RH, Roberts JE (2000) Photooxidation of lens alpha-crystallin by hypericin (active ingredient in St. John's wort). J Photochem Photobiol 72:200–203

Senol FS, Orhan IE, Ozgen U, Renda G, Bulut G, Guven L, Sekeroglu N (2016) Memory-vitalizing effect of twenty-five medicinal and edible plants and their isolated compounds. S Afr J Bot 102:102–109

Sezik E, Yesilada E, Honda G, Takaishi Y, Takeda Y, Tanaka T (2001) Traditional medicine in Turkey X. Folk medicine in Central Anatolia. J Ethnopharmacol 75:95–115

Shetty S, Udupa S, Udupa L (2008) Evaluation of antioxidant and wound healing effects of alcoholic and aqueous extract of *Ocimum sanctum* Linn in rats. Evid Based Complement Alternat Med 5:95–101

Smelcerovic A, Zuehlke S, Spiteller M, Raabe N, Özen T (2008) Phenolic constituents of 17 *Hypericum* species from Turkey. Biochem Syst Ecol 36:316–319

Sparenberg B, Dembch L, Hoelzl J (1993) Untersuchungen iiber antidepressive Wirkstoffe von Johanniskraut. Pharm Ztg Wiss 6(138):50–54

Tanker N (1971) Studies on *Hypericum scabrum* L. J Facult Pharma Ankara Univ 1:10–15

Tatli I, Sahpaz S, Kupeli AE, Martin-Nizard F, Gressier B, Ezer N, Bailleul F (2009) Antioxidant, anti-inflammatory and anti nociceptive activities of Turkish medicinal plants. Pharm Biol 47:916–921

Tekelova D, Repcak M, Zemkova E, Toth J (2000) Quantitative changes of dianthrones, hyperforin and flavonoids content in the flower ontogenesis of *Hypericum perforatum*. Planta Med 66:778–780

Turkoglu S, Turkoglu I, Celik S, Bahsi M, Gur S, Parlak AE (2015) Some biological compounds, antioxidant and antimicrobial activities of endemic *Hypericum uniglandulosum*. Chem Nat Compd 51:615–619

Tuzlaci E, Aymaz PE (2001) Turkish folk medicinal plants part IV: Gönen (Balıkesir). Fitoterapia 72:323–343

Unal E, Mavi A, Kara A, Cakir A, Sengul M, Yildirim A (2008) Antimicrobial and antioxidant activities of some plants used as remedies in Turkish traditional medicine. Pharm Biol 46:207–224

Upton R (Ed.) (1997) St. John's wort-*Hypericum perforatum*. American herbal Pharmacopoei and therapeutic compendium. Herbal gram (40) (summer), p 32

Uslusoy F, Nazıroğlu M, Çiğ B (2017a) Inhibition of the TRPM2 and TRPV1 channels through *Hypericum perforatum* in sciatic nerve injury-induced rats demonstrates their key role in apoptosis and mitochondrial oxidative stress of sciatic nerve and dorsal root ganglion. Front Physiol 8:335

Uslusoy F, Nazıroğlu M, Övey İS, Sönmez TT (2017b) *Hypericum perforatum* L. supplementation protects sciatic nerve injury induced apoptotic, inflammatory and oxidative damage to muscle, blood and brain in rats. J Pharm Pharmacol 71:83–92

Vural G (2008) Ethnobotanical features of some of the wild plants on the Honaz Mountain and its environment Ethnobotanic. Master Thesis. Afyon Kocatepe University, Graduate School of Natural and Applied, Sciences, Afyon

Weber ND, Murmy BK, North j A, Wood SG (1994) The antiviral agent hypericin has in vitro activity against HSV-1 through nonspecific association with viral and cellular membranes. Antivir Chem Chemother 5:83–90

Yesil Y (2007) An ethnobotanical study in Kürecik District (Malatya/Akçadag). Master Thesis. Istanbul University, Institute of Health Science, Department of Pharmaceutical Botany, Istanbul

Yesilada E, Honda G, Sezik E, Tabata M, Goto K, Ikeshiro Y (1993) Traditional medicine in Turkey IV. Folk medicine in the Mediterranean subdivision. J Ethnopharmacol 39:31–38

Yeşilada E, Honda G, Sezik E, Tabata M, Fujita T, Tanaka T, Taked Y, Takaishi Y (1995) Traditional medicine in Turkey. V Folk medicine in the inner Taurus Mountains. J Ethnopharmacol 46:133–152

5 Morphogenesis of Cotyledon Explants and Plantlet Regeneration of Tomato (*Lycopersicon esculentum* Mill.) Under Salt Stress

M. M. A. N. Shanika and Thayamini H. Seran

Abstract

This study was aimed to determine the responses of salt stress on cotyledon callus of tomato cv. KC-1 for plant regeneration. Cotyledon explants from in vitro raised seedlings were placed on MS medium containing 1.0 mg/L BAP and 0.2 mg/L NAA or 1.0 mg/L NAA to select a suitable medium for callus formation. After 6 days of culture, green calli were initiated on the cut surfaces of cotyledon explants. Callus formation was observed in 81.2% of the explants on the medium with 1.0 mg/L BAP and 1.0 mg/L NAA. After 2 weeks of culture, cotyledon produced calli were transferred to the medium containing 1.0 mg/L BAP plus 1.0 mg/L NAA without (control) or with NaCl concentrations (20, 40, and 60 mM). Callus was evaluated after 4 weeks of subculture (i.e., after 6 weeks of initial culture) under salt stress based on callus colour, fresh and dry weights of callus and chlorophyll content of callus. Results showed that the effects of tomato KC-1 cultivar on the tested parameters had a significant effect on different salinity levels. The calli that were placed on 60 mM NaCl level were highly affected when compared with control. Calli were then transferred to the medium with 2.0 mg/L BAP and 0.2 mg/L NAA for plant regeneration. Within 4 weeks, plantlets were formed on the control treatment and 20 mM salt stressed calli and they were acclimatized under ex vitro conditions.

Keywords

Callus · In vitro culture · Plantlet regeneration · Salt effect · Tomato

M. M. A. N. Shanika · T. H. Seran (✉)
Department of Crop Science, Faculty of Agriculture, Eastern University, Chenkalady, Sri Lanka
e-mail: thayaminis@esn.ac.lk

I. Siddique (ed.), *Propagation and Genetic Manipulation of Plants*,
https://doi.org/10.1007/978-981-15-7736-9_5

5.1 Introduction

Tomato (*Lycopersicon esculentum* Mill.) belongs to the family Solanaceae. It is important among the vegetable crops for human consumption and diverse uses in the world as it contains a considerable number of vitamins and minerals. Tomato products are carotenoid-rich foods. High dietary intake of tomato foodstuffs can reduce coronary heart disease in consequence of decreasing Low-density lipid (LDL) Cholesterol levels (Silaste et al. 2007). As a result, it is a market potential vegetable crop that is grown in the field, greenhouses and net houses not only in Sri Lanka but also in most of the countries (Bhatia et al. 2004). However, there are biotic and abiotic stresses to limit tomato production mainly under field conditions. Among them, salinity is one of the vital stresses in restrictive crop production (Mahajan and Tuteja 2005). In general, plant growth may be declined as a result of toxic salt ion accumulation in plants that can cause osmotic stress (Silveria et al. 2009). Sivritepe et al. (2003) reported that seedling growth at an early stage is more susceptible to salinity in many plants. Crop yields reduce in many plants when the electrical conductivity value of saturation soil extract increases above the threshold level value (Essington 2015). No significant yield reduction was revealed in most commercial cultivars of tomato grown in salinity level up to 2.5 dsm^{-1} (Singh et al. 2012). Tomato varieties namely KC-1, Tharindu, T-245, Marglobe, KWR, Roma, Thilina and Ravi are generally cultivated in Sri Lanka.

Biotechnological techniques have been applied to enhance the yield and quality of vegetable and other crops (Gosal et al. 2010; Ayobami et al. 2013). An effective protocol of in vitro culture technique is necessary for employing the benefits of plant tissue culture for genetic improvement of crops (Sheeja et al. 2004). Plant tissue culture technique has been effectively used to select tolerant cell lines against diverse biotic and abiotic stresses (Bhatia et al. 2004; Manoj et al. 2011). The establishment of callus is an important step for the in vitro culture studies. Exogenous plant growth regulators in culture medium play a vital role in callus induction and organ differentiation. According to a literature survey, IAA (Indole-3-acetic acid), NAA (-α-naphthalene acetic acid), 2,4-D (2,4-Dichlorophenoxy acetic acid), Zeatin and 6-BAP (6-benzyl amino purine) are commonly used in tissue cultures of tomato for callus formation and plant regeneration (Namitha and Negi 2013). This experiment was done to determine the effect of salt stress on callus formation from cotyledon explants of tomato cv. KC-1 for plant regeneration.

5.2 Materials and Methods

This research was done in 2015–2016 at the Tissue Culture Laboratory, Department of Crop Science, Faculty of Agriculture, Eastern University of Sri Lanka. Tomato seeds (KC-1 cultivar) used in this research work were obtained from the Horticultural Crop Research and Development Institute, Sri Lanka. The tomato seeds were dipped in 70% (v/v) ethyl alcohol for 1 min and then surface sterilized by using 20% (v/v) Clorox™ (5.25% sodium hypochlorite) with two drops of Tween 20 for

20 min. After that, they were washed four times with sterilized distilled water and dried on autoclaved filter paper for 15 min under sterile conditions. They were then inoculated on MS (Murashige and Skoog 1962) basal medium with 3% sucrose and 0.8% agar to germinate under in vitro conditions. In the present study, all cultures were kept at 25 ± 1 °C under white fluorescent light in a photoperiod of 16 h and 8 h dark. The light intensity of 2000–2500 lux and 70% relative humidity were maintained.

5.2.1 Callus Induction on Cotyledon Explants

In vitro raised seedlings (Fig. 5.1) were utilized to excise cotyledon explants. The cotyledon explants (0.5 cm long) were carefully detached from the seedlings under aseptic conditions subsequently they were placed (abaxial side) on MS basal medium containing 1.0 mg/L BAP and 0.2 mg/L NAA (Sherkar and Chavan 2014) or 1.0 mg/L BAP and 1.0 mg/L NAA to induce callus. The explants were observed at regular intervals for in vitro responses. This experiment was done twice and each replicate had 16 explants.

5.2.2 Salt Stress on Cotyledon Callus

From the previous work in this study, it was noted that the percentage of callogenesis was high in the cultured cotyledons on the culture medium with 1.0 mg/L BAP and 1.0 mg/L NAA. Hence, after 2 weeks of initial culture, the cotyledons producing calli on the above medium were placed on the MS media containing 1.0 mg/L BAP plus 1.0 mg/L NAA and also different concentrations of NaCl (0, 20, 40, 60 mM). The cultures were observed constantly for callus proliferation in each culture medium. After 4 weeks of culture in different salt concentrations, the colour of

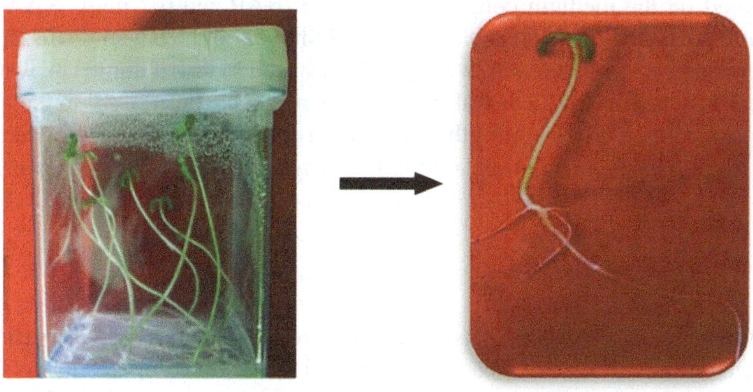

Fig. 5.1 In vitro raised 12 days old seedling

callus, fresh and dry weights of callus, water content % of callus and chlorophyll content of callus were assessed. This experiment was laid out in a completely randomized design with three replications and repeated twice.

5.2.3 Fresh and Dry Weights of Callus

After 4 weeks of culture, proliferating calli on the MS media with different NaCl concentrations were separately collected from the culture media and fresh weight of callus was measured in each salinity treatment. Thereafter, the collected callus was dried at 40 °C for 48 h and the constant weight of dried callus was determined. After measuring the fresh and dry weights of callus in each salt treatment, water content percentage of callus was estimated.

5.2.4 Chlorophyll Contents of Callus

After 4 weeks, 100 mg of calli were separately harvested from the media containing the different concentrations of NaCl. The collected calli from each treatment were placed individually in a glass vial (28 mL capacity). Subsequently, 10 mL of 80% acetone was added into each vial. Then, calli were blended with a homogenizer, and the extract was collected in a test tube to measure chlorophyll $_a$ (Chl_a) and chlorophyll $_b$ (Chl_b) contents using a UV visible spectrophotometer at 663 nm and 645 nm wavelengths respectively. The Chl_a, Chl_b and total Chl contents of calli were measured on a fresh basis using Arnon's equation (Arnon 1949).

5.2.5 Plant Regeneration from Cotyledon Callus

After 4 weeks of culture in different salt levels, the proliferating callus was subcultured on the medium containing 2.0 mg/L BAP and 0.2 mg/L NAA in line with the reports of Godishala et al. (2011) and Sherkar and Chavan (2014) for plant regeneration. In vitro plantlets were gently washed with running tap water and followed by sterile distilled water. They were then transferred to small pots containing soil and cow dung in the ratio of 1:1 (v/v). The pots were covered with a transparent polythene bag to retain high relative humidity around the plants.

5.2.6 Statistical Analysis

The data obtained in callus weights were subjected to analysis of variance using SAS 9.1.3 version. The treatment means of callus weights were compared using Tukey's Honestly Significant Difference Test at 5% significant level. A regression analysis was performed for chlorophyll contents of callus in different culture media and also correlation coefficients were calculated to correlate the chlorophyll content with NaCl concentration incorporated into the culture medium.

5.3 Results and Discussion

5.3.1 Callus Formation on Cotyledon Explants

Cotyledon explants from in vitro raised seedlings were placed on the medium with 1.0 mg/L BAP and 0.2 mg/L NAA or 1.0 mg/L NAA to select a suitable medium for callus formation. After 6 days of culture, yellowish-green calli were mostly initiated on the cut surfaces of cotyledon explants. Callus was formed in 81.2% of the cultured explants on the medium containing 1.0 mg/L BAP plus 1.0 mg/L NAA at 2 weeks of culture. Gerszberg et al. (2016) indicated that cotyledon is the more suitable explants to induce shoot for plant regeneration. Sherkar and Chavan (2014) observed a high percentage of callus formation in the cultured cotyledon explants on the media with 1.0–4.0 mg/L BAP and 0.2 mg/L NAA, however, shoot regeneration was not obtained. Properties of callus initiation in several explants, mainly associated with the quantity of endogenous plant growth regulators and their uptake, types of auxin and cytokinin used, as well as action mode of hormones (Gupta et al. 2010). The different types and concentrations of plant growth regulators have been used for in vitro plant regeneration of tomato. In vitro plantlets can be regenerated from callus through organogenesis or somatic embryogenesis procedure as stated by Verma et al. (2011) and other researchers.

5.3.2 Salinity Stress on Cotyledon Callus

After 2 weeks of culture, cotyledons produced calli were placed on the medium containing 1.0 mg/L BAP and 1.0 mg/L NAA without (control) or with NaCl concentrations (20, 40, and 60 mM) to assess their salt tolerance. After 4 weeks on salt media (i.e., after 6 weeks of initial culture), fresh and dry weights, water content and chlorophyll content of calli were measured. Moreover, the colour of callus in each medium was observed. The results showed that salt stress affected the tested parameters of callus derived from cotyledon of tomato KC-1 cultivar. Callus formation on different NaCl treatments after 4 weeks (after 6 weeks of initial culture) is shown in Fig. 5.2. Most of the researchers used callus to evaluate its physiological and biochemical responses to salt stress (Shibli et al. 2007; Ghane et al. 2014; Alharby et al. 2016).

5.3.3 Salt Stress on Colour of Callus

In this experiment, the colour of calli was observed at a regular interval. Callus colour was changed by increasing NaCl concentrations. After 2 weeks of culture,

Fig. 5.2 Effects of salt stress on cotyledon calli after 4 weeks of culture in NaCl treatments (**a–d**) (Note: Callus cultures on MS media with 1.0 mg/L BAP + 1.0 mg/L NAA supplemented with 0 mM (**a**) 20 mM (**b**) 40 mM (**c**) and 60 mM (**d**) NaCl concentrations)

yellowishgreen colour calli were gradually turned to greenish-brown colour in the medium with high salt concentration (60 mM). Yellowish-green to green colour was mostly observed on calli in the control treatment (0 mM NaCl) but the colour of callus was brownish-green in the 40 mM NaCl treatment after 4 weeks of culture. The colour change of calli with increasing salt content is in harmony with the studies in potato (Rahnama and Ebrahimzadeh 2004), sugarcane (Gandonou et al. 2006) and rice (Rattana and Bunnag 2015) response to salinity stress. Bouiamrine and Diouri (2012) stated that change in colour of cell mass to brownish colour was generally exploited as a sign of tissue culture in tolerance to osmotic stress.

5.3.4 Salt Stress on Weights of Callus

The results showed that after 4 weeks of culture, the fresh weight of callus in the control (0 mM NaCl) treatment was significantly ($P < 0.05$) higher (0.74 g) than that in salt-treated media (Fig. 5.3). The lowest fresh weight (0.28 g) of callus was recorded in the 60 mM NaCl concentration. Furthermore, it was noted that the fresh weight of callus was slowly decreased in the 40 mM NaCl as compared to the 20 mM NaCl concentration but there was no remarkable variation between them.

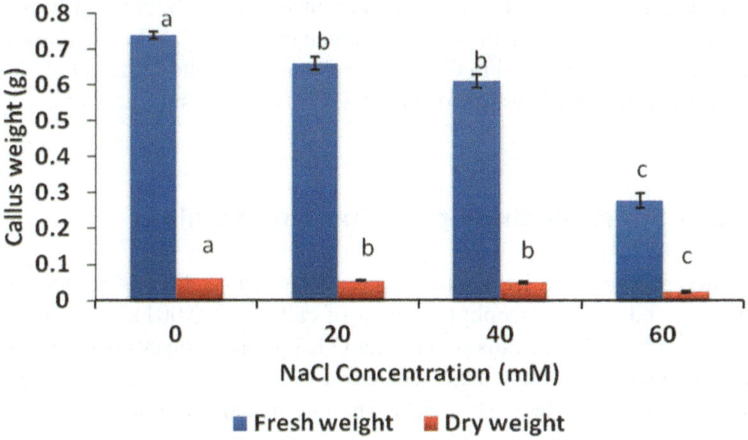

Fig. 5.3 Effect of different salinity levels on fresh and dry weights of cotyledon callus after 4 weeks of culture (Means followed by the same letter in each bar are not significantly different according to Tukey's test at 5% significant level)

In contrast, the fresh weight of callus was considerably declined on high NaCl concentration (60 mM NaCl). It may be due to salinity stress on water uptake for its growth. The dry weight of callus was also high in the control treatment with an average of 0.062 g while a dry weight of callus was low in the 60 mM concentration with an average of 0.023 g. No significant difference was noted on the 20 and 40 mM salt concentrations.

The results showed that the dry weight of calli on control treatment was considerably ($P < 0.05$) varied from those in the other treatments. It was significantly low value in 60 mM NaCl medium among the tested treatments. Water and nutrient contents are associated with the salt concentration in plant tissue (Martínez-Ballesta et al. 2006; Gao et al. 2013). Yunita et al. (2014) mentioned about the reduction of callus growth due to osmotic effects and also indicated that more energy is needed for its metabolism. This is in harmony with Abbas et al. (2012) revealed that callus growth declined at the highest salinity level and osmotic stress inhibits the absorption of mineral nutrients by the cells. Plant growth regulators in the cells take part in a vital role in regulating plant responses to abiotic stress (Upreti and Sharma 2016). As a result of osmotic stress, the addition of salt ions in the cell mass inhibits the synthesis of growth regulators that are necessary for cell differentiation and plant growth (Gupta and Huang 2014; Al-Hattab et al. 2015).

In the present study, water content % of callus in salt-treated media (0–60 mM NaCl) was about 92%. The results exhibited that salt stress was not significantly affected by water content of callus. High osmotic stress from high salt level controlled plant cells to uptake water and some mineral nutrients in the culture medium (Cicek and Cakirlar 2002). Al-Khayri and Al-Bahrany (2004) noted a progressive decline in callus growth as increasing water stress. Queiros et al. (2007) stated that the water content is reduced as a result of the high osmotic stress of the medium supplemented with high salt concentration. Salinity at higher levels has a result in plant metabolism (Hasanuzzaman et al. 2013). The decrease in water content of callus was observed in *Carthamus tinctorius* (Soheilikhah et al. 2013) and *Stylosanthes guianensis* (Veraplakorn et al. 2013) due to osmotic stress.

5.3.5 Salt Stress on Chlorophyll Content of Callus

The obtained results revealed that there was a significant relationship between NaCl concentration and total chlorophyll content of calli ($P < 0.001$). The value of total chlorophyll content of the callus was higher (31.8 μg/g) in the control treatment than other media under in vitro salt stress whereas it was lower (16.4 μg/g) in the medium containing 60 mM NaCl level (Fig. 5.4). The findings are supported by EL-Meleigy et al. (2004) who reported that chlorophyll contents in the entire tomato genotypes tested under in vitro conditions were declined due to salinity stress. The total chlorophyll contents were negatively correlated with NaCl levels in the culture media ($r = -0.942$). The reduction of leaf chlorophyll content under high salinity stress might be due to the activity of the chlorophyll degrading enzyme, chlorophyllase (Megdiche et al. 2007; Noreen and Ashraf 2009).

As compared to the control treatment, chlorophyll a and chlorophyll b in tomato calli were decreased due to salt stress in the present study. It was also noted that reduction percentage of both chlorophyll contents were increased with increasing NaCl concentrations added to the culture media. Moreover, the decline in chlorophyll a was lower than that in chlorophyll b content of callus cultured in 60 mM NaCl medium. This effect has also been stated by Delfine et al. (1999) in spinach and EL-Meleigy et al. (2004) in tomato. Aliu et al. (2015) mentioned that the highest strength of salinity generally reduced the contents of both chlorophyll a and b over the control. Screening of different genotypes for salt tolerance can be performed

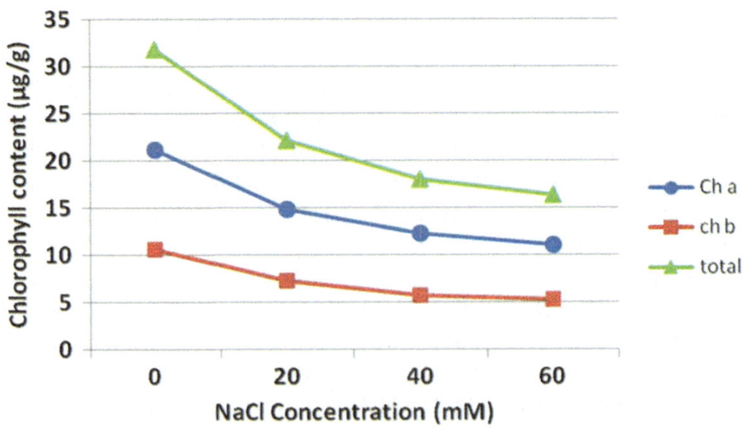

Fig. 5.4 Effect of NaCl concentrations on chlorophyll contents of calli

under in vitro conditions by using NaCl. In vitro screening of salt-tolerant genotypes has been reported in mulberry (Vijayan et al. 2003) and tomato (Rashed et al. 2016).

5.3.6 Plantlet Regeneration

In this experiment, the green nodules on the surface of cultured calli were noted after 1 week of culture. It was a sign of formation of embryoids like structures from the calli which was transferred to the medium with 2 mg/L BAP plus 0.2 mg/L NAA (Fig. 5.5). They were eventually developed into plantlets after 4 weeks. It was observed in calli under 0 mM and 20 mM salt stress. The nodule structures on the surface of the calli were also observed on 40 mM salt-stressed calli after 4 weeks. Somatic embryogenesis can be employed in mass plant propagation and transformation studies as stated by Yadav and Tyagi (2006). Godishala et al. (2011) stated that embryogenic callus was formed from cotyledon explants of tomato cultured on medium containing 2.0–4.0 mg/L BAP plus 0.2–0.5 mg/L IAA and also induction rate of somatic embryogenesis was highest on the medium with 3.5 mg/L BAP and

Fig. 5.5 Morphogenesis of 20 mM salt stressed callus cultured on MS medium containing 2.0 mg/L BAP plus 0.2 mg/L NAA after 1 week (**a**) and 2–3 weeks (**b–d**)

Fig. 5.6 Plantlets (**b**, **c**) regenerated from non-salt stressed callus (**a**) cultured on MS medium supplemented with 2.0 mg/L BAP plus 0.2 mg/L NAA after 4 weeks of culture

0.5 mg/L IAA. On the contrary, Durrani et al. (2017) informed that the calli exhibited greening but shoots did not appear on calli on the medium with 3 mg/L BAP and 0.3 mg/L NAA. In the present study, exogenous supply of BAP and NAA induced nodular structures on cotyledon callus for plant regeneration.

After 5 weeks, shoot elongation and root formation were observed on the medium with 2.0 mg/L BAP and 0.2 mg/L NAA (Fig. 5.6). Generally, a low concentration of BAP alone or in combination with auxin has been used for shoot elongation of plant species. But the relatively high concentration of BAP stimulated shoot bud development (Sathyagowri and Seran 2013; Lavakumaran and Seran 2014). Best callusing and shoot regeneration was noted from internode explants cultured in MS medium with 2.0 mg/L BAP and 0.2 NAA mg/L NAA (Sherkar and Chavan 2014; Shanika et al. 2018). Mohamed et al. (2010) reported that 2.0 mg/L BAP was better for growth traits of tomato among 0–4 mg/L concentrations of BAP. In the present study, the plantlets were collected from the in vitro culture and were then transferred to small pots containing a mixture of soil and cattle manure. Subsequently, they were acclimatized.

5.4 Conclusion

The cotyledons were responded well on MS medium fortified with 1.0 mg/L BAP plus 0.2 mg/L NAA. The present study revealed that by increasing the NaCl concentrations on the callus, the dry and fresh weights of callus and chlorophyll contents of callus was decreased, when cultured on the medium containing 1.0 mg/L BAP and 1.0 mg/L NAA. Tomato was more sensitive to salinity on callus formation with a significant reduction for salinities above 40 mM NaCl and a considerable reduction was observed on callus formation above this level. Formation of embryoids like structures was observed on the control treatment and 20 mM salt-stressed calli on the medium with 2.0 mg/L BAP and 0.2 mg/L NAA. In vitro plantlet formation was noticed on the same MS medium. This may lead to the mass production of plantlets within a short period. To increase the general tolerance mechanism against salt stress, future study is needed for the crop improvement program.

References

Abbas MF, Jasim AM, Al-Zubaidy BH (2012) The effect of proline on growth and ionic composition of embryogenic callus and somatic embryos of the date palm (*Phoenix dactylifera* L. CV. Ashkar) under NaCl stress. Int J Farming Allied Sci 1:82–87

Alharby HF, Metwali EMR, Fuller MP, Aldhebiani AY (2016) Impact of application of zinc oxide nanoparticles on callus induction, plant regeneration, element content and antioxidant enzyme activity in tomato (*Solanum lycopersicum* mill.) under salt stress. Arch Biol Sci 68:723–735

Al-Hattab ZN, Al-Ajeel SA, El-Kaaby EA (2015) Effect of salinity stress on *Capsicum annuum* callus growth, regeneration and callus content of capsaicin, phenylalanine, proline and ascorbic acid. J Life Sci 9:304–310

Aliu S, Rusinovci I, Fetahu S, Gashi B, Simeonovska E, Rozman L (2015) The effect of salt stress on the germination of maize (*Zea mays* L.) seeds and photosynthetic pigments. Acta Agricult Slovenica 105:85–94

Al-Khayri JM, Al-Bahrany AM (2004) Growth, water content, and proline accumulation in drought-stressed callus of date palm. Biol Plant 48:105–108

Arnon DI (1949) Copper enzymes in isolated chloroplasts polyphenoloxidase in *Beta vulgaris*. Plant Physiol 24:1–15

Ayobami AS, Valesca A, Vidal BF, Vasco A (2013) Biotechnology and agriculture. J Biosaf Health Educ 1:103

Bhatia P, Ashwath N, Senaratna T, Midmore D (2004) Tissue culture studies of tomato (*Lycopersicon esculentum*). Plant Cell Tissue Organ Cult 78:1–21

Bouiamrine E, Diouri M (2012) Response of durum wheat (*Triticum durum* Desf.) callus culture to osmosis induced drought stress caused by polyethylene glycol (PEG). Ann Biol Res 3:4555–4563

Cicek N, Cakirlar H (2002) The effect of salinity on some physiological parameters in two maize cultivars. Bulgar J Plant Physiol 28:66–74

Delfine S, Alvino A, Villani MC, Loreto F (1999) Restrictions to carbon dioxide conductance and photosynthesis in spinach leaves recovering from salt stress. Plant Physiol 119:1101–1106

Durrani NS, Ahmad D, Jalal A, Rajab H, Khan MS (2017) The effect of explant sources and growth regulators on callus induction and regeneration in different tomato cultivars. J Animal Plant Sci 27:481–489

El-Meleigy EA, Gabr MF, Mohamed FH, Ismail MA (2004) Responses to NaCl salinity of tomato cultivated and breeding lines differing in salt tolerance in callus cultures. Int J Agric Biol 6:19–26

Essington ME (2015) Soil and water chemistry: an integrative approach, 2nd edn. CRC Press, New York

Gandonou CB, Errabii T, Abrini J, Idaomar M, Senhaji NS (2006) Selection of callus cultures of sugarcane (*Saccharum* sp.) tolerant to NaCl and their response to salt stress. Plant Cell Tissue Organ Cult 87:9–16

Gao S, Liu KT, Chung TW, Chen F (2013) The effects of NaCl stress on *Jatropha* cotyledon growth and nitrogen metabolism. J Soil Sci Plant Nutr 13:99–113

Gerszberg A, Hnatuszko-Konka K, Kowalczyk T, Kononowicz AK (2016) Efficient in vitro callus induction and plant regeneration protocol for different polish tomato cultivars. Notul Bot Horti Agrobot Cluj-Napoca 44:452–458

Ghane SG, Lokhande VH, Nikam TD (2014) Growth, physiological, and biochemical responses in relation to salinity tolerance for in vitro selection in oil seed crop *Guizotia abyssinica* Cass. J Crop Sci Biotechnol 17:11–20

Godishala V, Mangamoori L, Nanna R (2011) Plant regeneration via somatic embryogenesis in cultivated tomato (*Solanum lycopersicum* L). J Cell Tissue Res 11:2521–2528

Gosal SS, Wani SH, Kang MS (2010) Biotechnology and crop improvement. J Crop Improv 24:153–217

Gupta B, Huang B (2014) Mechanism of salinity tolerance in plants: physiological, biochemical, and molecular characterization. Int J Genom 2014:1–18

Gupta P, Sharma S, Saxena S (2010) Callusing in *Stevia rebaudiana* (natural sweetener) for steviol glycoside production. Int J Agricult Biol Sci 1:30–34

Hasanuzzaman M, Nahar K, Fujita M (2013) Plant response to salt stress and role of exogenous protectants to mitigate salt-induced damages. In: Ahmad P, Azooz MM, Prasad MNV (eds) Ecophysiology and responses of plants under salt stress. Springer, New York, USA, pp 25–87

Lavakumaran L, Seran TH (2014) Effect of 6-benzyl-aminopurine and thidiazuron on in vitro shoot organogenesis of *Aloe vera* (L.) Burm. F. Chilean J Agricult Res 74:497–501

Mahajan S, Tuteja N (2005) Cold, salinity and drought stresses: an overview. Arch Biochem Biophys 444:139–158

Manoj KR, Rajwant KK, Rohats S, Manu PG, Dhawan AK (2011) Developing stress tolerant plants through in vitro selection- an overview of the recent progress. Environ Exp Bot 71:89–98

Martínez-Ballesta M d C, Silva C, Lopez-Berenguer C, Cabañero FJ, Carvajal M (2006) Plant aquaporins: new perspectives on water and nutrient uptake in saline environment. Plant Biol 8:535–546

Megdiche W, Ben Amor N, Debez A, Hessini K, Ksouri R, Zuily-Fodil Y, Abdelly C (2007) Salt tolerance of the annual halophyte *Cakile maritima* as affected by the provenance and the developmental stage. Acta Physiol Plant 29:375–384

Mohamed N, Ismail MR, Rahman MH (2010) In vitro response from cotyledon and hypocotyls explants in tomato by inducing 6-benzylaminopurine. Afr J Biotechnol 9:4802–4807

Murashige T, Skoog F (1962) A revised medium for rapid growth and bioassays with tobacco tissue cultures. Physiol Plant 15:473–497

Namitha KK, Negi P (2013) Morphogenetic potential of tomato (*Lycopersicon esculentum*) cv. Arka Ahuti to plant growth regulators. Notulae Sci Biol 5:220–225

Noreen Z, Ashraf M (2009) Changes in antioxidant enzymes and some key metabolites in some genetically diverse cultivars of radish (*Raphanus sativus* L.). Environ Exp Bot 67:395–402

Queiros F, Fidalgo F, Santos I, Salema R (2007) In vitro selection of salt tolerant cell lines in *Solanum tuberosum* L. Biol Plant 51:728–734

Rahnama H, Ebrahimzadeh H (2004) The effect of NaCl on proline accumulation in potato seedlings and calli. Acta Physiol Plant 26:263–270

Rashed MRU, Roy MR, Paul SK, Haque MM (2016) In vitro screening of salt tolerant genotypes in tomato (*Solanum lycopersicum* L.). J Horticult 3:186

Rattana K, Bunnag S (2015) Differential salinity tolerance in calli and shoots of four rice cultivars. Asian J Crop Sci 7:48–60

Sathyagowri S, Seran TH (2013) A note on morphological studies of shoot bud initiation from ginger (*Zingiber officinales* Rosc) rhizome cultured in vitro conditions. Sudan J Agricult Res 22:101–106

Shanika MMAN, Seran TH, Shiyamala B (2018) Plant regeneration from cotyledonary nodal explants of tomato (*Lycopersicon esculentum* mill.) cultivar KC-1. In: Proceedings of the 5th Ruhuna international science and technology conference held on 15th February 2018. University of Ruhuna, Sri Lanka, p 74

Sheeja TE, Mondal AB, Rathore RKS (2004) Efficient plantlet regeneration in tomato (*Lycoperiscon esculentum* a.B. mill.). Plant Tissue Cult 14:45–53

Sherkar HD, Chavan AM (2014) Studies on callus induction and shoot regeneration in tomato. Sci Res Rep 4:89–93

Shibli RA, Kushad M, Yousef GG, Lila MA (2007) Physiological and biochemical responses of tomato microshoots to induced salinity stress with associated ethylene accumulation. Plant Growth Regul 51:159–169

Silaste ML, Alfthan GAA, Kesäniemi YA, Hörkkö S (2007) Tomato juice decreases LDL cholesterol levels and increases LDL resistance to oxidation. Br J Nutr 98:1251–1258
Silveria JAG, Arujo SAM, Lima JPMS, Viegas RA (2009) Roots and leaves display contrasting osmotic adjustment mechanisms in response to NaCl salinity in *Atriplex nummularia*. Exp Environ Bot 66:1–8
Singh J, Sastry EVD, Singh V (2012) Effect of salinity on tomato *(Lycopersicon esculentum* mill.) during seed germination stage. Physiol Mol Biol Plants 18:45–50
Sivritepe N, Sivritepe HO, Erifl A (2003) The effects of NaCl priming on salt tolerance in melon seedlings grown under saline conditions. Sci Hortic 97:229–237
Soheilikhah Z, Karimi N, Ghasmpour HR, Zebarjadi AR (2013) Effects of saline and mannitol induced stress on some biochemical and physiological parameters of *Carthamus tinctorius* L. varieties callus cultures. Aust J Crop Sci 7:1866–1874
Upreti KK, Sharma M (2016) Role of plant growth regulators in abiotic stress tolerance. In: Rao NKS, Shivashankara KS, Laxman RH (eds) Abiotic stress physiology of horticultural crops. Springer, India, pp 19–46
Veraplakorn V, Nanakorn M, Kaveeta L, Suwanwong S, Bennett IJ (2013) Variation in ion accumulation as a measure of salt tolerance in seedling and callus of *Stylosanthes guianensis*. Theor Exp Plant Physiol 25:106–115
Verma SK, Yucesan BB, Gurel S, Gurel E (2011) Indirect somatic embryogenesis and shoot organogenesis from cotyledonary leaf segments of *Digitalis lamarckii* Ivan. an endemic medicinal species. Turk J Biol 35:743–750
Vijayan K, Chakraborti SP, Ghosh PD (2003) In vitro screening of mulberry (*Morus* spp.) for salinity tolerance. Plant Cell Rep 22:350–357
Yadav PR, Tyagi R (2006) Biotechnology of plant tissue. Discovery Publishing House, New Delhi, India
Yunita R, Khumaida N, Sopandie D, Mariska I (2014) Growth and regeneration of rice (*Oryza sativa* L.) callus in salt medium. Biosci Res 11:4–9

Callus Culture Systems in *Salix* L.: The Limited Database 6

Afsheen Shahid and Mohammad Anis

Abstract

Salix L. is the largest genera of family Salicaceae. Commonly known as 'willows', the genus *Salix* is a large contributor to the energy crop systems, an important source of natural 'aspirin' (Salicin and derivatives) and a large number of active constituents with medicinal properties. The dioecious nature of the tree and different geographical distributions of the male and female counterparts renders it difficult to propagate via seeds. Several studies have been published on the clonal micropropagation of *Salix* L. in the past 40 years; however, only four studies reported the callus culture systems. The presented review focus on the limited knowledge of callus induction protocols in *Salix* L. throughout the genera and the importance of pursuing this line of study.

Keywords

Micropropagation · Willow · Medicinal Plants · Salicaceae

6.1 Introduction

'Plant tissue culture is a tool that is an adjunct to studies on whole plants'—Donald K Dougall.

A. Shahid (✉)
Chair of Tree Physiology, Institute of Forest-Sciences, Albert-Lüdwigs University, Freiburg, Germany

M. Anis
Department of Botany, Aligarh Muslim University, Aligarh, India

I. Siddique (ed.), *Propagation and Genetic Manipulation of Plants*,
https://doi.org/10.1007/978-981-15-7736-9_6

Plant tissue culture systems, either for the purpose of clonal propagation or in vitro secondary metabolite production, play a pivotal role in the area of conservation biology. Tissue culture systems provide a fighting chance for the dwindling biodiversity, making it possible to upscale the clonal plantlet production for endangered and threatened plant species. Another important aspect of tissue culture is saving the medicinally important plant species from overexploitation and endangerment. The tissue culture protocols have been established by various working groups for almost all the medicinally important plant families, some getting more attention than the other. Induction and establishment of plant callus cultures with the aim of secondary metabolite accumulation depend on various factors.

Callus culture systems have been designed and established in a large number of plants. The early tissue culture strategies were all focused on cell elongation, division and dedifferentiation to achieve a free growing cell mass to be directed in desired line of indirect plant regeneration, somatic embryogenesis, de-novo metabolite production and elicitation of pre-existing active constituents. Callus masses are the raw material for various established and emerging tissue culture branches.

> 'Despite advances in the field of organic chemistry, plants are still an important commercial source of chemical and medicinal compounds'—Dodds and Roberts (1985)

Plants produce a vast array of primary and secondary product with secondary biosynthetic pathways branching at different developmental stages of the plant tissue and organs. Some secondary products are of universal occurrence throughout the plant kingdom while some can be specific to a particular plant family, genus, species or sometimes the natural intraspecific hybrids. Within the single plant species, the synthesis, distribution and accumulation pattern of one secondary metabolite may also vary. Isolated cell systems and callus culture systems provide a working base to study the differential response of specific plant organs from different species, genus and families for secondary metabolite production, biotransformation and even finding novel more potent substitutions of pre-established active constituents from various plant species. Callus culture forms the basis of all the advanced research, be it the genetic manipulation, metabolic engineering, biotransformation, metabolite production or secondary embryogenesis/adventitious regeneration.

6.1.1 *Salix* L. (The Willow)

Salix L. is the type genus and by far the largest genera of family Salicaceae (Azuma et al. 2000), which includes a total of 55 genera in the family and 772 recognized species (Table 6.1). Still, about one-third of the genetic biodiversity of the family is unrecognized and unpublished. Salix L. includes approximately 377 species (The Plant List 2013) widely distributed in the northern hemisphere including subtropical, temperate and cold regions (Chen et al. 2010) and Indo-Malesia and mountains of China in the Indian subcontinent. It is known to be a taxonomically difficult genus.

Table 6.1 List of genera in family Salicaceae including number of species listed

S. No.	Genera	No. of species
1	*Abatia*	8
2	*Aberia*	1
3	*Ahernia*	1
4	*Aigeiros*	1
5	*Azara*	9
6	*Banara*	17
7	*Bartholomaea*	1
8	*Bembicia*	1
9	*Bennettiodendron*	1
10	*Bivinia*	1
11	*Calantica*	1
12	*Carrierea*	2
13	*Casearia*	64
14	*Chosenia*	1
15	*Craepaloprumnon*	1
16	*Dovyalis*	9
17	*Euceraea*	3
18	*Flacourtia*	7
19	*Gossypiospermum*	2
20	*Guidonia*	8
21	*Hasseltia*	2
22	*Hasseltiopsis*	1
23	*Hecatostemon*	1
24	*Homalium*	33
25	*Idesia*	1
26	*Itoa*	1
27	*Laetia*	9
28	*Lasiochlamys*	1
29	*Ludia*	4
30	*Lunania*	2
31	*Macrohasseltia*	1
32	*Macrothumia*	1
33	*Neopringlea*	2
34	*Neoptychocarpus*	2
35	*Neosprucea*	4
36	*Olmediella*	1
37	*Oncoba*	14
38	*Phyllobotryon*	1
39	*Pineda*	1
40	*Pleiarina*	1
41	*Pleuranthodendron*	4
42	*Poliothyrsis*	1
43	*Populus*	87

(continued)

Table 6.1 (continued)

S. No.	Genera	No. of species
44	*Prockia*	3
45	*Pseudoscolopia*	1
46	*Ryania*	6
47	***Salix***	**377**
48	*Samyda*	6
49	*Scolopia*	9
50	*Tetrathylacium*	2
51	*Tisonia*	6
52	*Toisusu*	1
53	*Trimeria*	1
54	*Xylosma*	45
55	*Zuelania*	1

The infrageneric classification of *Salix* L. is highly controversial (Chen et al. 2010) even though several classifications have already been proposed and modified on the basis of morphology, phenology and molecular taxonomy for it (Chao et al. 1998; Leskinen and Alström-Rapaport 1999; Skvortsov 1999; Ohashi 2001). The application of *Salix* L. as the main source of herbal NSAIDs (non-steroidal anti-inflammatory drugs) in the medicine industry makes it an interesting choice for regulated secondary metabolite production in vitro along with micropropagation strategies for clonal propagation and conservation of elite genotypes and phenotypes.

Salix L. is a very diverse and useful genus with ecologically and medicinally important species. There are several species of *Salix* L. which are listed (IUCN Red List) as endangered (*S. kusani*), vulnerable (*S. xanthicola, S. megnifica, S. floridana, S. tortosina,* etc.) or near –threatened (*S. libani, S. tirrenico, S. crataegifolia, S. canareinsis, S. ionica,* etc.) all over the globe. Still, there are handful of reports available for conservation, sustainable biochemical utilization (in vitro metabolite production) and replantation of the species (via clonal propagation) using tissue culture methods.

6.1.2 Tissue Culture in *Salix*

Since the beginning of plant tissue culture methods, various workers studied the different aspects of tissue culture in *Salix* L. All the studies conducted in the area have been summarized in Table 6.2 and can be broadly categorized in two lines of investigation.

6.1.2.1 Micropropagation/Direct Organogenesis In Vitro

Various species of *Salix* L. have been selected by tissue culturists for micropropagation studies (Table 6.2). The earliest successful attempts for direct organogenesis were recorded by Bhojwani (1980) using nodal explants of hybrid

Table 6.2 Tissue culture studies on various species in genus *Salix*, for the past four decades

S. no.	Species	Tissue culture strategy	References
1	*S. matsudana X alba (NZ1002)*	Direct organogenesis	Bhojwani (1980)
2	*S. alba* *S. fragilis* *S. viminalis*	Direct organogenesis	Chalupa (1983)
3	*S. babylonica*	Direct shoot regeneration	Dhir et al. (1984)
4	*S. schwerinii* *S. Dasyclados* *S. caprea* *S. viminalis* *S. caprea (hybrid)*	Direct shoot regeneration	Bergman et al. (1985)
5	*S. viminalis*	**Callus culture and somatic embryogenesis**	Grönroos et al. (1989)
6	*S. exigua*	**Callus cultures**	Stoehr et al. (1989)
6	*S. viminalis*	**Protoplast culture**	Vahala and Eriksson (1991)
7	*S. caprea*	Direct shoot regeneration	Neuner and Beiderbeck (1993)
8	*S. fragilis X lispoclados*	Ovary culture	Agrawal and Gebhardt (1994)
9	*S. tarraconensis*	Direct shoot regeneration	Amo-Marco and Lledo (1996)
10	*S. caprea*	Direct shoot regeneration	Liesebach and Naujoks (2004)
11	*S. nigra*	Direct adventitious shoot regeneration	Lyyra et al. (2006)
12	*S. pseudolariogyne*	Direct regeneration	Park et al. (2008)
13	*S. tetrasperma*	Direct regeneration	Khan et al. (2011); Khan and Anis (2012)
14	*S. viminalis* *S. cinerea*	**Callus culture**	Skálová et al. (2012)
15	*S. viminalis*	Direct regeneration	Regueira et al. (2018)
16	*S. tetrasperma*	**Callus cultures**	Shahid and Anis (2018)
17	*S. tetrasperma*	Synthetic seeds	Khan et al. (2018)
18	*S. retusa* *S. alpina*	Direct shoot multiplication	Chornobrov et al. (2019)

willow (*Salix matsudana X S. alba* NZ-1002), while Chalupa (1983) successfully reported the direct organogenesis in three different species of *Salix* (*S. alba, S. fragilis, S. viminalis*) from nodal segments and axillary bud induction and proliferation in vitro. Later in the decade, Dhir et al. (1984) reported successful shoot proliferation under the effects of potential cytokinin, that is, BAP followed by NAA supported rooting and later acclimatization of the plantlets. Following the example of these early successful tissue culture studies, Bergman et al. (1985)

attempted direct shoot regeneration via axillary bud induction in five different species of *Salix* (*S. schwerinii, S. dasyclados, S. caprea, S. viminalis, S. caprea*) under the influence of BAP augmentation.

Salix caprea (Goat willow) is the most studied species of *Salix* for tissue culture and micropropagation. After Bergman et al. (1985), Neuner and Beiderbeck (1993) and Liesebach and Naujoks (2004) also reported the direct shoot regeneration from nodal segments of *S. caprea*.

Agrawal and Gebhardt (1994) took advantage of the in vitro regeneration strategies to establish the intrageneric hybrid between *S. fragilis* (female) X *S. lispoclados* (male) using ovary culture. The hybrid seedlings were grown in vitro and shoot multiplication was achieved using the nodal segments from them. This study revealed the positive outcomes of Woody Plant Medium augmented with balanced auxin to cytokinin ratio. In 1996, Amo-Marco and Lledo used in vitro direct shoot regeneration for the clonal propagation and conservation of *S. tarraconensis*, an endemic and threatened species of *Salix*. Another important species of *Salix* is *S. nigra* (Black willow) which is the largest and the most commercially important species of willows in North America. Lyyra et al. (2006), using immature inflorescence explants, successfully implemented the direct adventitious shoot formation. This only reported study of tissue culture studies in *S. nigra*. Khan et al. (2011) and Khan and Anis (2012) extensively studied the in vitro regeneration methods for *S. tetrasperma* (Indian willow) using different explants and an array of phytohormone and growth media combinations. For *S. tetrasperma*, a conservation protocol was established by Khan et al. (2018) using in vitro synthetic seeds made of aseptic nodal segments instead of somatic embryos. This study utilized the micropropagation and regeneration protocols established in previous studies by Khan et al. (2011) and Khan and Anis (2012). In the most recent study, Regueira et al. (2018) used liquid immersion techniques for enhanced in vitro multiplication and acclimatization of *S. viminalis* axillary shoots successfully. An incomplete tissue culture study was published by Chornobrov et al. (2019) on direct shoot regeneration of two endangered *Salix* species, *S. retusa* and *S. alpina*. The protocol was incomplete due to the lack of rooting and acclimatization of multiplied in vitro axillary shoots.

6.1.2.2 Callus Cultures

However, a large number of species have been selected for direct micropropagation studies in vitro (Table 6.2), the number of species studied for callus cultures are limited and the successful callus culture establishment in the genus are even lesser. There are very few reports available on successful callus induction in genus *Salix* L. In most instances, the utilization of tissues from both vegetative and reproductive origin was studied and compared for their callus induction and successful proliferation capabilities. The first attempts have been made by Grönroos et al. (1989) for the purpose of indirect organogenesis. They successfully utilized isolated pistils from female catkins of *S. viminalis* for callus culture production followed by somatic embryogenesis. Later in the same year, Stoehr et al. (1989) utilized developing leaves of mature *S. exigua* for successful callus culture establishment. The leaf explants were inoculated on WPM medium supplemented with high auxin and low cytokinin. However, they reported efficient callus growth but after 7 weeks of incubation, the study indicated the slow growth rate of callus cultures. In the early 1990s, Vahala and Eriksson (1991) studied the protoplast isolation and regeneration from the suspension cultures in *S. viminalis* and *S. shwerinii* using the callus culture protocol from Grönroos et al. (1989); however, the protoplast recovery for fusion or genetic transformation and re-establishment of viable callus cultures was unsatisfactory. They observed the formation of several microcalli, but without the regeneration potential. Skálová et al. (2012) reported successful callus cultures in few species of *Salix*. The group studied various species of *Salix* for the purpose of direct and indirect regeneration of plant tissue from vegetative as well as reproductive organs. The direct organogenesis was successful in most cases, but the tissue culture of reproductive organs showed minimal results. Callus cultures were established from *S. viminalis* leaves as well as isolated anthers, anthers of *S. caprea* and microcalli were observed from ovule cultures of *S. viminalis* and leaf culture of *S. cinerea*. They used combinations of various basal medium, growth regulators and additives for various regeneration studies. For callus culture, induction-modified MS medium augmented with NAA and BA was reported to be most beneficial, but the callus induction efficiency was found to be insignificant. The most recent callus establishment was reported by Shahid and Anis (2018), using inflorescence explants of *Salix tetrasperma* Roxb., also known as 'Indian Willow', to establish non-organogenic callus cultures for secondary metabolite enhancement, in particular, anthocyanin production and elicitation. In this study, MS medium augmented with 2,4-D (1.0 μM) was reported to be optimum for maximum callus proliferation efficiency.

Over an expanse of four decades, only five studies were published for callus cultures in few species of genus *Salix* and four-fifth of these were focused on revival of the differentiated state of the tissue and plant regeneration either via shoot regeneration or somatic embryogenesis and only one-fifth on the secondary metabolite production and enhancement.

6.1.3 Future Strategies

The first requirement for genetic manipulation of a plant species in vitro is an efficient callus induction protocol, followed by regeneration of plants from callus. Biotransformation and enhanced/novel metabolite production also require successful and reproducible callus culture protocols. *Salix* L., the type genus and largest genera of family Salicaceae, lacks both the direct and indirect regeneration protocols in vitro in more than 80% of the species. Very few species have been studied so far and little accomplishments have been made towards novel biotechnological studies in the genera.

Salix L. is an important genus of family Salicaceae and includes a large number of species. It is a potential source of 'salicin' and salicylate derivatives. Till date the genus has been ignored by the scientific community as an alternative herbal source of 'aspirin' – a potent anti-inflammatory drug over chemically synthesized drug.

More focus is demanded in future, to establish an herbal drug line derived from the plant. Various enhancement and biotransformation studies need to be conducted for effective industrial production of the herbal active constituents from *Salix* L., which all need the base of successful and reproducible tissue culture protocols. Establishment of callus culture systems is a prerequisite to initiate an industrial-scale suspension culture system. Various species of the genus need to be screened for in vitro callus culture establishment and the upregulation of 'salicin' and other salicylate derivatives.

References

Agrawal DC, Gebhardt K (1994) Rapid micropropagation of hybrid willow (*Salix*) established by ovary culture. J Plant Physiol 143:763–765

Amo-Marco JB, Lledo MD (1996) In vitro propagation of *Salix tarraconensis* Pau ex font Quer, an endemic and threatened plant. In Vitro Cell Develop Biol Plant 32:42–46

Azuma T, Kajita T, Yokoyama J, Ohashi H (2000) Phylogenetic relationships of *Salix* (Salicaceae) based on rbcL sequence data. Am J Bot 87:67–75

Bergman L, Von Arnold S, Eriksson T (1985) Effects of N6-benzyladenine on shoots of five willow clones (*Salix* spp.) cultured in vitro. Plant Cell Tissue Org Cult 4:135–144

Bhojwani SS (1980) Micropropagation method for a hybrid willow (*Salix matsudana*× *alba* NZ-1002). N Z J Bot 18:209–214

Chalupa V (1983) In vitro propagation of willows (*Salix* spp.), European mountain-ash (*Sorbus aucuparia* L.) and black locust (*Robinia pseudoacacia* L.). Biol Plant 25:305–307

Chao N, Gong GT, Liu J (1998) On the classification and distribution of the family Salicaceae. J Sichuan Forest Sci Technol 19:9–20

Chen JH, Sun H, Wen J, Yang YP (2010) Molecular phylogeny of *Salix* L. (Salicaceae) inferred from three chloroplast datasets and its systematic implications. Taxon 59:29–37

Chornobrov O, Bilous S, Chornobrov O, Manko M (2019) Peculiarities of morphogenesis of the endangered species of willow (*Salix* spp.) in vitro. Biol Ther Dent 65:48–55

Dhir KK, Angrish R, Bajaj M (1984) Micropropagation of *Salix babylonica* through in vitro shoot proliferation. Proc Plant Sci 93:655–660

Dodds JH, Roberts LW (1985) Experiments in plant tissue culture, 2nd edn. Cambridge University Press, New York, p 232

Grönroos L, Von Arnold S, Eriksson T (1989) Callus production and somatic embryogenesis from floral explants of basket willow (*Salix viminalis* L.). J Plant Physiol 134:558–566

Khan MI, Anis M (2012) Modulation of in vitro morphogenesis in nodal segments of *Salix tetrasperma* Roxb. Through the use of TDZ, different media types and culture regimes. Agrofor Syst 86:95–103

Khan MI, Ahmad N, Anis M (2011) The role of cytokinins on in vitro shoot production in *Salix tetrasperma* Roxb.: a tree of ecological importance. Trees 25:577–584

Khan MI, Ahmad N, Anis M, Alatar AA, Faisal M (2018) In vitro conservation strategies for the Indian willow (*Salix tetrasperma* Roxb.), a vulnerable tree species via propagation through synthetic seeds. Biocatal Agric Biotechnol 16:17–21

Leskinen E, Alström-Rapaport C (1999) Molecular phylogeny of Salicaceae and closely related Flacourtiaceae: evidence from 5.8 S, ITS 1 and ITS 2 of the rDNA. Plant Syst Evol 215:209–227

Liesebach M, Naujoks G (2004) Approaches on vegetative propagation of difficult-to-root *Salix caprea*. Plant Cell Tissue Organ Cult 79:239–247

Lyyra S, Lima A, Merkle SA (2006) In vitro regeneration of *Salix nigra* from adventitious shoots. Tree Physiol 26:969–975

Neuner H, Beiderbeck R (1993) In vitro propagation of *Salix caprea* L. by single node explants. Silvae Genetica 42:308–308

Ohashi H (2001) Salicaceae of Japan. In: Scientific report. Toboku University, 4th Series Biology, vol 40. University of Michigan, Michigan, pp 269–396

Park SY, Kim YW, Moon HK, Murthy HN, Choi YH, Cho HM (2008) Micropropagation of *Salix pseudolasiogyne* from nodal explants. Plant Cell Tissue Organ Cult 93:341

Regueira M, Rial E, Blanco B, Bogo B, Aldrey A, Correa B, Varas E, Sánchez C, Vidal N (2018) Micropropagation of axillary shoots of *Salix viminalis* using a temporary immersion system. Trees 32:61–71

Shahid A, Anis M (2018) Production and enhancement of anthocyanin pigment in callus cultures of Salix tetrasperma Roxb. J Plant Sci Res 34(1):9–14

Skálová D, Navrátilová B, Richterová L, Knit M, Sochor M, Vašut R (2012) Biotechnological methods of in vitro propagation in willows (*Salix* spp.). Open Life Sci 7:931–940

Skvortsov AK (1999) Willows of Russia and adjacent countries. Taxonomical and geographical revision. (English translation of Skvortsov 1968. Ivy SSSR. Sistematicheskiy i geograficheskiy Obzor [willows of the USSR. Taxonomic and geographic revision.]. Nauka, Moscow. 262 p.). Univ. Joensuu. Faculty of Mathematics and Natural Sciences, Joensuu, Finland

Stoehr MU, Cai M, Zsuffa L (1989) In vitro plant regeneration via callus culture of mature *Salix exigua*. Can J For Res 19:1634–1638

The Plant List (2013) Version 1.1 Published on the Internet; http://www.theplantlist.org/

Vahala T, Eriksson T (1991) Callus production from willow (*Salix viminalis* L.) protoplasts. Plant Cell Tissue Organ Cult 27:243–248

An Efficient Plant Regeneration System Via Leaf Derived Callus of *Solanum melongena* L.

7

Hina Khan

Abstract

An efficient plant regeneration protocol developed of economically important vegetable crop, *Solanum melongena* L., through leaf-derived callus. Leaf explant from 20 days old aseptic seedling used for callus induction on MS (Murashige and Skoog) medium augmented with different concentration of 2,4-D and 2,4,5-T. A light green friable callus from leaf explant was obtained in 7.5 μM (2,4-D) which did not have regenerative potential. Differentiation of adventitious shoot was obtained on culturing the friable callus on MS medium augmented with various concentration of BA and Kinetin alone or in combination. The maximum number of adventitious shoots (65.5 ± 2.02) was obtained from leaf-derived callus on MS basal medium augmented with 2.0 μM BA and 0.1 μM kinetin. The potentiality of differentiation of nodular callus was influenced through callus age. For the induction of roots, the regenerated shoots cultured on MS medium supplemented with 0.5 μM Indole-3-acetic acid (IAA) or on hormone-free MS medium. The in vitro regenerated plantlets were acclimatized and shifted into the field. The in vitro regenerated plantlets were comparable to the parent plants. The developed plant regeneration protocol is a valuable tool in biotechnology for the improvement of eggplant crops in the plant breeding system.

Keywords

Adventitious shoots · Eggplant · Friable callus · Plant regeneration · *Solanum melongena*

H. Khan (✉)
Department of Botany and Environmental Science, Sri Guru Granth Sahib World University, Fatehgarh Sahib, Punjab, India

I. Siddique (ed.), *Propagation and Genetic Manipulation of Plants*,
https://doi.org/10.1007/978-981-15-7736-9_7

7.1 Introduction

Eggplant (*Solanum melongena* L.) is a dominant agronomically crop and cultivated worldwide and provided nutritive as well as medicinal values. The highest production of Solasodine was detected in the young fruits of *S. melongena* (Prajapati et al. 2004) that is the important alternative source of steroidal drug production in pharmaceutical industry. Biotechnology techniques are most important for plant improvement requires the establishment of an efficient improved protocol for plant regeneration. Several reports on somatic embryogenesis and micropropagation reported in eggplant (Kamat and Rao 1978; Matsuoka and Hinata 1979). It is possible that plant tissue culture techniques may add to the solution of several agronomic problems in this crop. Plant regeneration from callus cultures has been reported from a variety of explants in eggplant (Alicchio et al. 1982; Swamy et al. 1988; Rotino et al. 1992; Jyoti et al. 1994; Miyoshi 1996; Franklin et al. 2004; Hossan et al. 2007; Ray and Hassan 2011; Roumiana et al. 2012). Since the information on organogenesis through the process of differentiation of the callus from different explants is not sufficient at hand, the present studies were attempted with the aim of inducing high frequency plant regeneration through de-differentiated callus tissues for its improvement.

7.2 Materials and Methods

7.2.1 In Vitro Seedlings

S. melongena L. seeds (cv. Pusa Purple Round) were soaked overnight and seeds were kept under tap water for half an hour to remove sand particles. Under the laminar air flow, the washed seeds were sequentially decontaminated with 5% (v/v) Teepol (10 min) and 0.1% (w/v) $HgCl_2$ (5 min) and finally five times rinsed with double distilled water. To optimize seedling, the disinfected seeds were cultured on Murashige and Skoog (1962) medium. For callus induction and differentiation, cotyledonary leaves were excised from aseptic seedling and cut up into smaller segments (1–1.5 cm) and cultured on MS medium containing different concentration of auxins 2,4-D, 2,4,5-T and NAA.

7.2.2 Culture Medium and Conditions

Murashige and Skoog (MS) medium contained vitamins, sucrose 3% (w/v) and agar 8% (w/v). The pH of media was fixed at 5.8 with 0.1 N NaOH before autoclaving at 121 °C for 20 min. The culture tubes were kept in the culturing room under appropriate temperature (25 ± 2 °C), photoperiod (16-h), illumination (50 μ mol m^{-2} $s^{-1)}$ provides with fluorescent tubes.

7.2.3 Production of Callus

For callus induction, MS medium alone or with different concentrations (5, 10, 15, 20 and 25 μM) of each 2,4-D, 2,4,5-T and NAA were used. Four weeks after incubation, well-developed callus production was recorded as the percentage of callus induction from leaf explants.

7.2.4 Shoot Differentiation

Light green friable callus was subcultured on MS basal medium (control medium) or MS medium augmented with BA and Kinetin alone or in combination (0.5, 1, 2 and 3 μM) for differentiation of shoot buds. The mean number of shoots and length of shoots induced from the nodular callus were evaluated after 4 weeks in culture.

7.2.5 In Vitro Rooting

For the development of roots from regenerated shoots, single shoot was excised and cultured on hormone-free MS medium or with different concentration of IAA (0.5, 1, 1.5, 2 and 2.5 μM). Data were noted later 3 weeks of culturing of shoots as percentage of rooting from the regenerated shoots, mean number and length of roots.

7.2.6 Acclimatization

For the hardening, the plantlets were excised from the rooting medium and washed with tap water and transferred to paper pots containing sterilized soilrite and covered with perforated polythene bags to maintain high humidity and kept in culturing room and irrigated at 2 days intervals with ½ strength liquid MS medium and bags were removed subsequently later 4 weeks. The saplings were shifted to the earthen pots comprising soil and sustained in green house.

7.2.7 Data Analysis

All the experimentations were repeated twice and 12 tubes were used for every single treatment. The influence of diverse treatments used in each experiment was measured and data were statistically evaluated via SPSS 10 (SPSS Inc., Chicago, USA). Duncan's test at $P = 0.05$ used for the analysis of range values among means.

7.3 Results and Discussion

Differentiation of callus was observed from the margin of leaf on MS medium contained with 10 μM 2,4-D within 15 days of culture. It was observed that the entire leaf segments transformed into callus in 4 weeks (Table 7.1, Fig. 7.1a). Initiation of callus was not perceived on MS (control medium). The light green, friable callus was ineffectual for persuading morphogenesis. Similar results were observed in *S. melongena* L. (Gleddie et al. 1985; Nigra et al. 1989). On medium containing 2,4,5-T generally the least callus formation and growth was observed, while media containing NAA induced callus at low frequency and grew relatively slow with the formation of roots as consistent by Alicchio et al. (1982). Differentiation of adventitious shoots from the friable callus achieved on MS medium augmented with BA and Kinetin alone or in combination. The production of adventitious shoots from the nodular callus on MS medium supplemented with Kinetin was lesser as compared to BA (Table 7.2). Similar morphogenic changes where BA was more effective than kinetin for shoot proliferation have been reported by Gleddie et al. (1985). Highest number of shoots obtained on media comprising 2.0 μM BA and 1.0 μM kinetin (Fig. 7.1b), where the shoot mean number from callus (72.3 ± 2.02) and mean shoot length (4.3 ± 0.29) was noted (Table 7.2, Fig. 7.1c). Superiority of BA and Kinetin combination for induction of shoot has been reported in *Capsicum annuum* (Soniya and Nair 2004). The number of shoots per culture was 2–72 from leaf-derived calluses, respectively (Table 7.2). The different response could be due to the different concentrations of auxin and cytokinins used in the MS medium (Saxena et al. 1997; Patra et al. 1998; Faisal and Anis 2003, 2005). The regenerated adventitious shoots inoculated on hormone-free MS medium or with different concentration of IAA (0.5, 1, 1.5, 2 and 2.5 μM) for induction of roots. Maximum number of roots (8.3 ± 0.23) from regenerated

Table 7.1 Effectiveness of auxins on inducing of callus from leaf explants after 4 weeks in culture of *S. melongena*

		% Response
Auxin	Concentration (μM)	Leaf
2,4-D	0	0
	5	64
	10	100
	15	87
	20	77
	25	65
2,4, 5-T	5	28
	10	50
	15	52
	20	54
	25	64
NAA	5	35
	10	30
	15	24
	20	43
	25	36

Fig. 7.1 Induction of callus, shoot regeneration and complete plant establishment from leaf explant of *Solanum melongena* L. (**a**) Callus induction from leaf explant on MS + 2,4-D (10 μM). (**b**) Shoot differentiation from organogenic callus on MS + BA (2.0 μM) + Kinetin (1.0 μM). (**c**) Elongation of shoots on MS. (**d**) Rooting of in vitro regenerated shoot on MS medium. (**e**, **f**) A plant acclimatized to greenhouse conditions

Table 7.2 Effect of BA and Kinetin for shoot regeneration from leaf-derived callus of *S. melongena* after 4 weeks of culture

Treatments (μM)		Leaf explants	
BA	Kinetin	Mean number of shoots/callus	Mean shoot length (cm)
0	0	0	0
0.5	0	$5.3 \pm 0.02^{e,f}$	1.5 ± 0.15^{g}
1.0	0	$6.6 \pm 0.33^{e,f}$	$2.0 \pm 0.19^{d,e,f}$
2.0	0	15.0 ± 0.17^{d}	3.4 ± 0.21^{b}
3.0	0	8.4 ± 0.41^{e}	$1.6 \pm 0.28^{f,g}$
0	0.5	2.1 ± 0.16^{f}	$2.3 \pm 0.11^{d,e}$
0	1.0	2.2 ± 0.20^{f}	1.3 ± 0.17^{g}
0	2.0	2.6 ± 0.32^{f}	$1.8 \pm 0.15^{e,f,g}$
0	3.0	3.5 ± 0.32^{f}	$1.6 \pm 0.12^{f,g}$
2.0	0.5	21.6 ± 0.76^{c}	$2.5 \pm 0.26^{c,d}$
2.0	1.0	72.3 ± 2.02^{a}	4.3 ± 0.29^{a}
2.0	2.0	35.3 ± 1.19^{b}	$3.1 \pm 0.20^{b,c}$
2.0	3.0	33.2 ± 1.01^{b}	$2.3 \pm 0.17^{d,e}$

Mean values signify means ± SE. In a column means represent by the identical letter are not considerably different via the Duncan's multiple range test at $P = 0.05$

Table 7.3 Influence of control medium (MS) and IAA for production of roots from regenerated shoots of *S. melongena* after 3 weeks of culture

Treatments (μM)	% Rooting	Mean no. of roots/shoot	Mean root length (cm)
MS	90	8.3 ± 0.23^{a}	3.6 ± 0.24^{a}
MS + IAA (0.5)	50	4.6 ± 0.20^{b}	3.3 ± 0.28^{a}
MS + IAA (1.0)	35	3.6 ± 0.17^{c}	3.0 ± 0.17^{b}
MS + IAA (1.5)	30	2.5 ± 0.20^{d}	2.9 ± 0.18^{b}
MS + IAA (2.0)	20	2.5 ± 0.20^{d}	2.5 ± 0.20^{c}
MS + IAA (2.5)	15	2.0 ± 0.28^{e}	2.1 ± 0.17^{c}

Mean values signify means ± SE. In a column means represent the identical letter are not considerably different via the Duncan's multiple range test at $P = 0.05$

shoot and highest rooting percentage (90%) were obtained on hormone-free MS medium within 3 weeks of culture. Roots were prominent with several root hairs (Table 7.3, Fig. 7.1d). These observations indicate that a particular level of auxin was indeed necessary to induce rooting which was readily available endogenously (Minocha 1987). Although IAA induced rooting, percentage of rooting at 0.5 μM IAA was comparatively low (50%) and roots were thin and delicate. Similar observations were reported in *S. melongena* (Jyoti et al. 1994).

Rooted plantlets were transferred into plastic pots comprising sterilized soilrite for acclimatization (Fig. 7.1e, f). After transplantation in the greenhouse about 85% of plants survived where they attained normal flowering and fruiting.

7.4 Conclusion

In conclusion, *S. melongena* L. used as a model plant due to its high morphogenetic potentiality in tissue culture and useful for the establishment of efficient plant regeneration from leaf-derived callus provides platform for the improvement of crop varieties through biotechnological approaches and somaclonal variations.

References

Alicchio R, Grosso ED, Boschieri E (1982) Tissue cultures and plant regeneration from different explants in six cultivars of *Solanum melongena*. Experientia 38:449–450

Faisal M, Anis M (2003) Rapid mass propagation of *Tylophora indica* Merill via leaf callus culture. Plant Cell Tissue Org Cult 75:125–129

Faisal M, Anis M (2005) An efficient in vitro method for mass propagation of *Tylophora indica*. Biol Plant 49:257–260

Franklin G, Sheeba CJ, Sita GL (2004) Regeneration of eggplant (*Solanum melongena* L.) from root explants. In Vitro Cell Develop Biol Plant 40:188–191

Gleddie S, Keller WA, Setterfield G (1985) Plant regeneration from tissue, cell and protoplast cultures of several wild *Solanum* species. J Plant Physiol 119:405–418

Hossan MJ, Rehman M, Bari MA (2007) Establishment of cell suspension culture and plant regeneration of Brinjal (*Solanum melongena* L). J Plant Sci 2:407–413

Jyoti M, Babber S, Varghese TM (1994) In vitro regeneration of *Solanum melongena* L. Var. H-7. J Ind Bot Soc 73:339–340

Kamat MG, Rao PS (1978) Vegetative multiplication of eggplant (*Solanum melongena*) using tissue culture techniques. Plant Sci Lett 13:57–65

Matsuoka H, Hinata K (1979) NAA—induced organogenesis and embryogenesis in hypocotyl callus of *Solanum melongena* L. J Exp Bot 30:363–370

Minocha SC (1987) Plant growth regulators and morphogenesis in cell and tissue culture in forest trees. In: Bonga JM, Durjan DJ (eds) Cell and tissue culture in forestry, vol 1. Martinus Nijhoff, Publishers, Dordrecht, pp 50–66

Miyoshi K (1996) Callus induction and plantlet formation through culture of isolated microspores of eggplant (*Solanum melongena* L). Plant Cell Rep 15:391–395

Murashige T, Skoog F (1962) A revised medium for rapid growth and bioassays with tobacco tissue cultures. Physiol Plant 15:473–497

Nigra HM, Alvarez MA, Giulietti AM (1989) The influence of auxins, light and cell differentiation on solasodine production by *Solanum eleagnifolium* Cav. Calli. Plant Cell Rep 8:230–233

Patra A, Rai B, Rout GR, Das P (1998) Successful plant regeneration from callus cultures of *Centella asiatica* (Linn.) urban. Plant Growth Regul 24:13–16

Prajapati ND, Purohit SS, Sharma AK, Kumar T (2004) A hand book of medicinal plants. CRC Press, Boca Raton, FL, p 480

Ray BP, Hassan NKM (2011) In vitro regeneration of Brinjal (*Solanum melongena* L.) Bangla. J Agric Res 36:397–409

Rotino GL, Arpaia S, Iannacone R, Iannamico V, Mennella G, Onofaro V, Perrone D, Sunseri F, Xike Q, Sponga F (1992) *Agrobacterium* mediated transformation of *Solanum* spp. using *Bacillus thuringiensis* (Bt) gene effective against coleopterans. Proc. VIIIth meeting on genetics and breeding of *Capsicum* and eggplant, Rome, Italy, 7-10 September, Capsicum News Latter, special issue, 295–300

Roumiana VV, Kraptchev B, Stoeva D (2012) Indirect shoot organogenesis of eggplant (*Solanum melongena* L.). J Cent Eur Agric 13:446–457

Saxena C, Palai SK, Samantaray S, Rout GR, Das P (1997) Plant regeneration from callus cultures of *Psoralea corylifolia* Linn. Plant Growth Regul 22:13–17

Soniya EV, Nair GM (2004) Multiple shoot regeneration and indirect organogenesis in chilli pepper (*Capsicum annuum L*). Capsicum Eggplant Newslett 23:93–96
Swamy SM, Christopher T, Subhash K (1988) Multiple shoot formation in embryo culture of *Solamun melongena*. Curr Sci 57:197–198

Selection of Elite Genotype of a Multipurpose Forest Tree (*Pterocarpus marsupium* Roxb.) from Naturally Grown Populations in Central India

8

Anees Ahmad, Iram Siddique, and Ankita Varshney

Abstract

In the present study, a total of 50 trees of *Pterocarpus marsupium* were phenotypically identified on the basis of vitality and growth form of stems at ten population sites of different forest regions of central India. Planting materials (fruits/seeds) were collected and aseptically cultured on optimized ½ MS (Murashige and Skoog, Physiol Plant. 15:473–497, 1962) medium in combination with 0.5 μM GA_3 (gibberellic acid) for 21 days and data on in vitro seed germination characteristics were interpreted statistically. Selection of elite genotype of *P. marsupium* was accomplished by the integration of both biotechnological and statistical approaches. Based on the parameters evaluated, the region of central India, Anuppur (MAA) accessions were found to be fast-growing as it showed maximum values of in vitro germination parameters such as percentage germination (PG, 91.3%), speed germination (SG, 29.0 seeds day^{-1}), coefficient germination (CG, 8.23%), mean germination time (MGT, 12.15 days), seedling height (SH, 10.31 cm), vigor index (VI, 941.7), seedling fresh weight (SFW, 4.77 g per 10 seedlings), seedling dry weight (SDW, 0.90 g per 10 seedlings), chlorophyll content (2.03 mg g^{-1} FW of *Chl* a), 0.84 mg g^{-1} FW of *Chl* b, total chlorophyll (2.87 mg g^{-1} FW of T*Chl*) and carotenoids (0.70 mg g^{-1} FW of *Car*), after 21 days of seed germination. Whereas Chhindwara (MCD) accessions were found to be the slowest growing compared to others and exhibited minimum values, that is, 50.3% of PG, 9.7 seeds day^{-1} of SG, 7.37% of CG, 13.56 days of MGT, 6.27 cm of SH, 315.8 of VI, 3.15 g per ten seedlings of SFW, 0.31 g per ten seedlings of SDW, 1.20 mg g^{-1} FW of *Chl* a, 0.47 mg g^{-1} FW of *Chl* b, 1.65 mg g^{-1} FW of T*Chl* and 0.37 mg g^{-1} FW of *Car*, after 21 days of seed germination. The present study demonstrated the successful selection of healthy vegetation which could be used as a planting source material for developing

A. Ahmad (✉) · I. Siddique · A. Varshney
Plant Biotechnology Laboratory, Department of Botany, Aligarh Muslim University, Aligarh, India

I. Siddique (ed.), *Propagation and Genetic Manipulation of Plants*,
https://doi.org/10.1007/978-981-15-7736-9_8

in vitro propagation techniques for obtaining large scale young plantlets, which can be reintroduced in the wild within a comprehensive conservation project.

Keywords

Conservation · Gibberellic acid · In vitro germination · Percentage variation · Woody legume

Abbreviations

B_5	Gamborg medium
Car	Carotenoids
CG	Coefficient germination
Chl	Chlorophyll
GA_3	Gibberellic acid
MGT	Mean germination time
MS	Murashige and Skoog medium
PG	Percentage germination
PGR	Plant growth regulator
PV	Percentage variation
SDW	Seedling dry weight
SFW	Seedling fresh weight
SG	Speed germination
SH	Seedling height
VI	Vigor index
WPM	Woody plant medium

8.1 Introduction

Forest trees provide valuable resources to humankind for economical, environmental, and industrial development, as the human life is sustained directly and indirectly by these resources, supplying a wide range of goods and ecological services essential for survival and prosperity. Medicinal trees are major resources of traditional medicine system and pharmaceutical industries. They provide a big source of revenue and health security of poor communities in developing countries. *Pterocarpus marsupium* Roxb. is one of the most valuable multipurpose forest tree of India commonly known as "Indian Kino or Bijasal." The tree species has received a lot of attention in recent years in experimental as well as conservational studies because it has several evidences of having potential bioactivity. It is used to cure several diseases in traditional ayurvedic and unani medicine system for many centuries (Maurya et al. 2004; Mohire et al. 2007). It also possesses several pharmacological activities like antidiabetic (Dhanabal et al. 2006), antimicrobial and antioxidant (Chakraborty et al. 2010), cardiotonic (Mohire et al. 2007), anti-inflammatory (Sanders et al. 2005), antioxidant and analgesic (Tippani et al. 2010), anticancer activity (Remsberg et al. 2008), and cyclooxygenase (COX-2) inhibition activity (Hougee et al. 2005). Besides, the National Medicinal Plant Board (NMPB 2018) of India estimated the annual trade value of this tree as 200–-500 metric tonnes/year in India, and each 10–15 years old mature tree produces about 0.5–0.6 tonne of dry heartwood which cost around US $1200–1500 in international trade market. Its heartwood can be utilized for making herbal drug products as well as bridge, boat, and small-scale construction materials (Barstow 2017).

The only source of propagation of the species is seeds contained in winged fruits/pod whose germination is very low due to hard fruit coat, short viability, poor pod set, and pathogenic infection (Anis et al. 2005; Ahmad et al. 2018). Furthermore, species have been overexploited due to its economic trade use by pharmaceutical and herbal drug industries, local tribes for medicine and timber, and this has resulted in its fast disappearance from natural stands (Chand and Singh 2004; Husain et al. 2010). Therefore, in view of restricted distribution, uncontrolled exploitation, and its inherent qualities, there is an urgent need to conserve this valuable forest tree through various biotechnological methods for large-scale plantlets production. However, before developing any propagation protocol one most important aspect is the identification of healthy and elite vegetation of targeted species. For this purpose, we have chosen phenotypically healthy population patches from different geo-climatic region of central India based on their abundance in particular area with straight bole and disease-free.

In the present study, attempts have been made to identify elite genotype among the ten naturally grown populations in the central India by comparing various parameters such as (a) in vitro seed germination such as percentage germination (PG), speed germination (SG), coefficient germination (CG), and mean germination time (MGT), (b) seedling growth like seedling height (SH), vigor index (VI), seedling fresh (SFW), and dry weight (SDW), (c) photosynthetic pigments content, that is, chlorophyll (*Chl* a/b), total chlorophyll (T*Chl*), and carotenoids (*Car*) in in vitro germinated seedlings. The study may help in the identification and selection of superior planting materials prior to the development of an in vitro propagation protocol to be used in tree improvement program.

8.2 Materials and Methods

8.2.1 Study Areas and Plant Material Collection

The winged fruits (Fig. 8.2a) of *P. marsupium* were collected from ten naturally grown forest populations of central India (Table 8.1) between February and March 2016. Each population has more than five tree accessions (or according to availability) and distance between subsequent collected accessions within population was approximately more than 200 m. About 500 g were collected from each tree accession and transported to laboratory in polybags at room temperature, where seeds (Fig. 8.2b) were mechanically excised from the fruits with the help of stripper

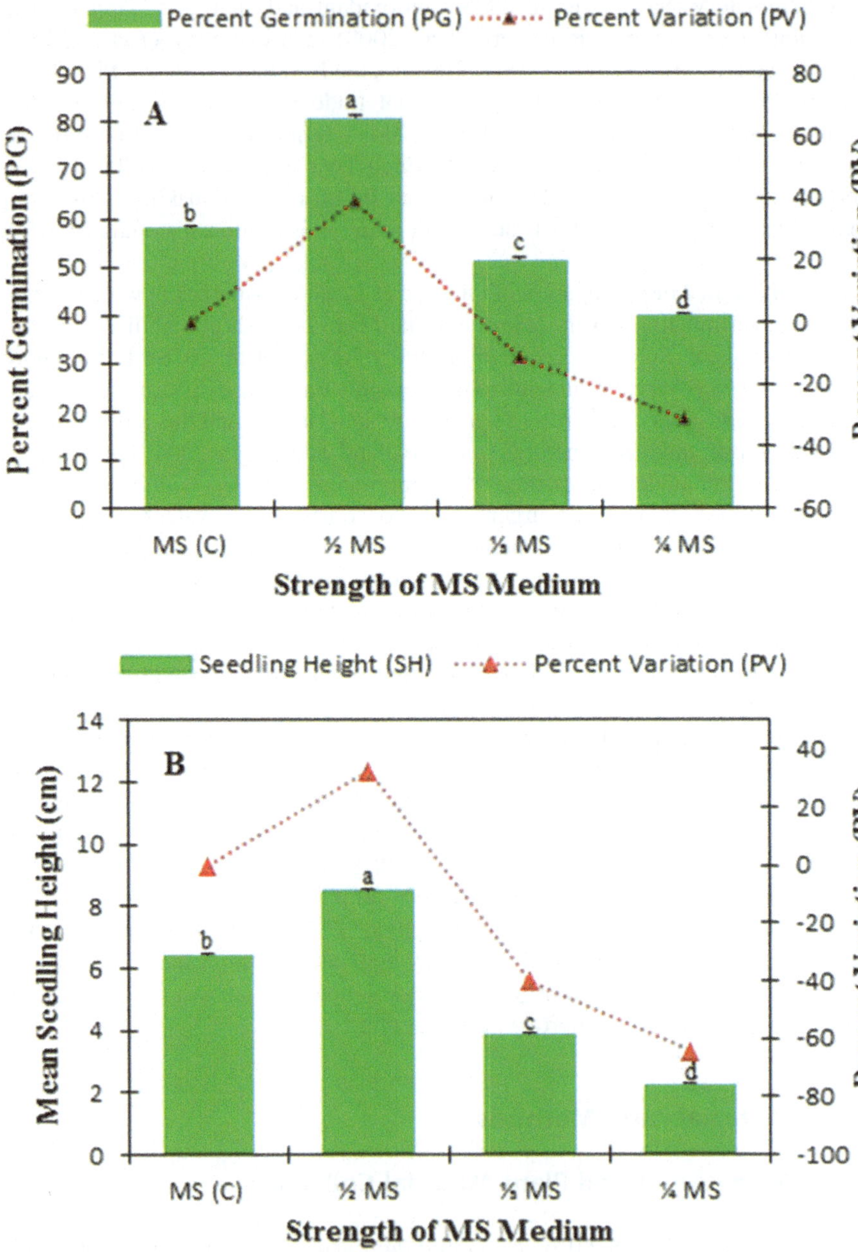

Fig. 8.1 Effect of various strength of MS medium augmented with GA_3 (0.50 μM), after 21 days of seed germination. (**a**) Percentage response of seed germination (**b**) seedling height. Full strength MS medium with optimum dose of GA_3 was considered as control and percentage variation (*PV*) of each treatment was calculated with reference to control (*C*). Bars represent the mean ± SE. Bars denoted by the same letter within response variables are not significantly different ($p = 0.05$) using Duncan's multiple range test

Fig. 8.2 In vitro germination of *Pterocarpus marsupium* seeds collected from Raigarh (CRK) forest region. (**a**) Winged fruits; (**b**) excised seeds; (**c**) 21-day-old seedlings germinated on ½ MS + 0.50 μM GA_3

cutter and thoroughly mixed together in a single lot to obtain compound samples of each population and stored at room temperature until further study.

8.2.2 Establishment of Aseptic Seed Culture Conditions

Surface sterilization was carried out using a two-step procedure. In the first step, seeds were washed thoroughly in running tap water for 15 min, immersed in 1% (w/v) Bavistin® solution for 8 min followed by washing with running tap water for 10 min and later treated with 5% (v/v) Labolene (Qualigens, India), a detergent for 10 min to remove adherent particles from the surface. Thereafter, seeds were kept in sterilized distilled water for imbibition for 24 h. In the second step, the seeds were sterilized with 0.1% (w/v) $HgCl_2$ for 4 min, followed by repeated washes with sterilized distilled water. Disinfected seeds were aseptically inoculated on jam bottles (Vista Biocell, India) or culture tube (25 × 150 mm, Borosil, India) containing 50 and 20 mL sterile culture medium, respectively. The media fortified with 3% (w/v) sucrose (Thermo Fisher Scientific India) as a carbon source, 0.75%

Table 8.1 Details of naturally grown populations of *Pterocarpus marsupium* Roxb. in different forest regions of the Central India

State	Populations				Geographical position		Forest type
	Forest region		Code	Mean GBH (cm)	Latitude (N)	Longitude (E)	
Madhya Pradesh	1	Mandla	MMK	125.0	22° 55′ 58″	80° 12′ 53″	Dry deciduous
	2	Jabalpur	MJH	104.0	22° 50′ 52″	79° 46′ 16″	Dry deciduous
	3	Anuppur	MAA	120.0	22° 40′ 31″	81° 43′ 32″	Semidry deciduous
	4	Hoshangabad	MHP	130.0	22° 39′ 23″	78° 23′ 03″	Dry deciduous
	5	Betule	MBG	128.0	22° 11′ 18″	77° 29′ 01″	Dry deciduous
	6	Seoni	MSS	122.0	21° 53′ 15″	79° 31′ 48″	Dry deciduous
	7	Chhindwara	MCD	138.0	21° 52′ 35″	78° 43′ 45″	Dry deciduous
	8	Balaghat	MBS	132.0	21° 51′ 46″	80° 21′ 01″	Dry deciduous
Chhattisgarh	9	Surguja	CSA	133.0	23° 03′ 48″	83° 17′ 25″	Dry deciduous
	10	Raigarh	CRK	128.0	22° 23′ 25″	84° 21′ 44″	Semidry deciduous

Where *GBH* Girth at breast height

(w/v) agar (Thermo Fisher Scientific India) as a gelling agent. The pH of media was adjusted to 5.8 with 1 N NaOH prior to autoclaving at 121 °C and 1.06 kg cm^{-2} pressure for 19 min. All culture tubes were incubated at 24 ± 2 °C under 16/8 h (light/dark) cycle with a photosynthetic photon flux density (PPFD) of 50 μmol m^{-2} s^{-1} provided by cool white fluorescent tubes (Philips, India) with 55 ± 5% relative humidity, regularly checked by thermohygrometer (Testo, India Pvt. Ltd.).

8.2.3 Culture Media for Seed Germination and Seedling Growth

Different basal media such as MS, WPM (Lloyd and McCown 1980), and B_5 (Gamborg et al. 1968) singly or in combination with GA_3 at different concentrations (0.1, 0.25, 0.5, 1.0, or 2.0 μM) were examined to detect the best medium where maximum seed germination response was obtained. Various strength (full, ½, $^1/_3$, and ¼) of optimized medium in combination with optimum dose of GA_3 were also evaluated for further improvement in in vitro seed germination. The MS medium without GA_3 was considered as control treatment for calculating percentage variation (PV) of each treatment. Hundred seeds of each treatment and five seeds per jam bottle were inoculated. All experiments were repeated three times. Data on percentage germination and plantlet height were recorded after 21 days of seed culture.

8.2.4 Selection of Elite Genotype

Identification of elite genotype of *P. marsupium* was accomplished by the integration of statistical and biotechnological approaches. The surface-sterilized seeds of ten collected populations were aseptically cultured on optimized seed germination ½ MS medium containing 0.5 μM GA_3. Performance of in vitro seed germination characteristics such as in vitro seed germination, seedling growth, and seedlings photosynthetic pigment contents of ten populations was considered as biological markers. Hundred seeds of each population were cultured in three repeated experiments. All the germination parameters were recorded at 3 days intervals and continued up to 21 days (3, 6, 9, 12, 15, 18, or 21 days; Fig. 8.3). Data on germination characteristics were interpreted statistically after 21 days. The slow-growing population was considered as a control population and percentage variation (PV) of each population was calculated with reference to control. The selection of superior population was done by following parameters using different statistical expressions:

8.2.4.1 Parameters Related to In Vitro Seed Germination

(a) Percentage germination (PG): Percentage germination was calculated according to ISTA (1999) using following formula

Fig. 8.3 In vitro germination of *Pterocarpus marsupium* seeds collected from Amarkantak (MAA) forest region on ½ MS medium supplemented with GA_3 (0.50 μM) at different days

$$PG = \frac{\text{Number of seeds germinated}}{\text{Total number of seeds cultured}} \times 100.$$

(b) Speed germination (SG): It was calculated by the using formula given by Maguire (1962).

$$SG = n_1/d_1 + n_2 - n_1/d_2 + \cdots n_x - (n_x - 1)/d_x,$$

where, n = number of germinated seeds, d = number of days, x = number of days to final count.

(c) Coefficient germination (CG): It was calculated by the following formula given by Copeland (1976).

$$CG = \frac{n_1 + n_2 + \cdots n_x}{n_1 d_1 + n_2 d_2 + \cdots n_x d_x} \times 100,$$

where, n = number of germinated seeds, d = number of days corresponding to n, x = number of days to final count.

(d) Mean germination time (MGT): It was calculated according to Ellis and Roberts (1981) formula.

$$MGT = \sum fx / \sum f,$$

where, f = seeds germinated on day x.

8.2.4.2 Parameters Related to Seedling Growth

(a) Seedling height (SH): Seedling height (cm) was measured after 21 days of seed culture. The mean height of 10 randomly selected seedlings of each population was recorded.

(b) Vigor index (VI): The value of vigor index was intended according to Abdul-Baki and Anderson (1973) by multiplying germination percentage and average height of ten randomly selected seedling of each population after 21 days of seed culture.

$$VI = PG \times \text{Mean seedling height}.$$

(c) Biomass content: mean fresh weight of ten randomly selected seedling of each population was recorded and then kept in shade for 24 h followed by oven drying at 85 °C for 48 h. Thereafter, dried seedlings were weighed to estimate the dry matter production, and the mean values were expressed in gram (g) per ten seedlings.

8.2.4.3 Parameters Related to Photosynthetic Pigment Content

The chlorophyll (*Chl* a/b) and carotenoids (*Car*) pigments were extracted from 21-day-old leaf tissues of ten randomly selected seedling of each population, following the MacKinney (1941) method. About 0.2 g fresh leaves were grounded in 80% (v/v) acetone solution with the help of mortar and pestle and filtered with filter paper (Whatman No. 1). The obtained extract was diluted to a final volume of 10 mL and quantification was performed by spectrophotometer (US-1700 Pharma Spec, Shimadzu, Kyoto, Japan). Optical density (OD) for chlorophyll solution was read at 645, 663 nm and for carotenoids at 480 and 510 nm wavelengths, and calculated using standard formulae. These pigments were expressed as milligram per gram of fresh weight (mg g^{-1} FW). Assessment of photosynthetic pigments and spectrophotometrical readings was repeated thrice. The experiments were conducted three times and data recorded at zero days considered as control for calculating percentage variation.

8.2.5 Statistical Analysis

All experiments were based on randomized block design and repeated three times with ten replicate for each treatment. The data on various parameters were subjected to one-way analysis of variance (ANOVA) using SPSS version 16 (SPSS Inc., Chicago, USA). The significance of differences among means was calculated using Duncan's multiple range tests at $P = 0.05$, and results were expressed as the mean with values parenthesized as percentage variations (PV). The PV was calculated by the following formula

$$\mathrm{PV} = \frac{\mathrm{Treatment} - \mathrm{Control}}{\mathrm{Control}} \times 100.$$

8.3 Results

8.3.1 Optimization of In Vitro Seed Germination Medium

Different basal media (MS, WPM, or B_5) single or in combination with various concentrations of GA_3 was examined for inducing maximum percentage of seed germination and seedling growth of randomly selected CRK (Raigarh) accessions. It showed maximum 58.3% (55.1% over control) seed germination with an average seedling height of 6.43 cm (106.0% over control) recorded on full strength MS containing 0.50 μM GA_3 after 21 days of seed culture (Table 8.2). After selection of appropriate basal media, various strengths of MS medium were also evaluated in combination with optimum concentration of GA_3 (0.50 μM) for further improvement in seed germination parameters. Of these, half-strength MS medium in combination with optimum GA_3 was found most suitable where a maximum 81% (38.9% over

Table 8.2 Effect of different media alone or in combination with GA_3 on seed germination and seedling height of *P. marsupium* after 21 days of seed culture

		Percentage germination		Seedling height (cm)	
Media	GA_3 (μM)	Mean ± SE	(PV)	Mean ± SE	(PV)
MS (C)	0.00	37.6 ± 0.57	$(0.0)^{f}$	3.12 ± 0.02	$(0.0)^{g}$
MS	0.10	40.0 ± 0.69	$(6.5)^{e}$	4.84 ± 0.03	$(55.0)^{c}$
MS	0.25	52.1 ± 0.15	$(38.5)^{b}$	5.39 ± 0.06	$(72.9)^{b}$
MS	0.50	58.3 ± 0.06	$(55.1)^{a}$	6.43 ± 0.06	$(106.0)^{a}$
MS	1.00	46.5 ± 0.12	$(23.7)^{c}$	4.95 ± 0.06	$(58.8)^{c}$
MS	2.00	41.7 ± 0.12	$(10.8)^{d}$	3.74 ± 0.04	$(19.8)^{e}$
WPM	0.00	25.3 ± 0.38	$(-32.6)^{l}$	2.05 ± 0.06	$(-34.4)^{l}$
WPM	0.10	28.4 ± 0.46	$(-24.6)^{j}$	2.56 ± 0.05	$(-17.8)^{j}$
WPM	0.25	31.1 ± 0.15	$(-17.4)^{h}$	3.75 ± 0.07	$(20.1)^{e}$
WPM	0.50	34.3 ± 0.06	$(-8.8)^{g}$	4.25 ± 0.04	$(36.3)^{d}$
WPM	1.00	29.4 ± 0.06	$(-21.8)^{i}$	3.78 ± 0.03	$(21.2)^{e}$
WPM	2.00	26.9 ± 0.12	$(-28.5)^{k}$	3.45 ± 0.04	$(10.7)^{f}$
B_5	0.00	16.0 ± 0.17	$(-57.4)^{o}$	1.38 ± 0.02	$(-55.8)^{m}$
B_5	0.10	20.1 ± 0.37	$(-46.5)^{n}$	2.65 ± 0.02	$(-15.1)^{i,j}$
B_5	0.25	24.7 ± 0.23	$(-34.4)^{l}$	2.93 ± 0.05	$(-6.0)^{h}$
B_5	0.50	27.2 ± 0.23	$(-27.6)^{k}$	3.38 ± 0.04	$(8.3)^{f}$
B_5	1.00	21.0 ± 0.12	$(-44.1)^{m}$	2.76 ± 0.06	$(-11.6)^{i}$
B_5	2.00	19.4 ± 0.12	$(-48.5)^{n}$	2.24 ± 0.05	$(-28.3)^{k}$

Values in parentheses are percentage variation (*PV*). Means values ($n = 30$) followed by the same letter with in column of each treatment are not significantly different ($p = 0.05$) using Duncan's multiple test. (*C*, control)

control) of seed germination (Fig. 8.1a) with an average seedling height of 8.52 cm (32.6% over control) (Fig. 8.1b) was recorded on randomly selected CRK accessions, after 21 days of seed culture (Fig. 8.2c).

8.3.2 In Vitro Seed Germination Parameters

Seeds of ten populations were aseptically cultured on ½ MS medium containing GA_3 (0.50 μM) and data were calculated on in vitro germination, after 21 days of culture. On the basis of germination parameters, MAA population was found to be fast-growing population compared to the others as it showed significantly increased values of germination parameters such as PG of 91.3% (81.5% over control), SG of 29.0 seeds day^{-1} (199.5% over control), and CG of 8.23% (11.6% over control), after 21 days of seed culture (Table 8.3). Also, MAA population showed the lowest value of MGT as 12.15 days (−10.4% over control) than others while MCD population indicates maximum value of MGT on 13.56 days after 21 days of culture (Table 8.3). These values represent fast- and slow-spreading capacity of populations, respectively (Table 8.3). On the other hand, MCD population was found to be slowest growing population compared to others and exhibited minimum values, that is, PG as 50.3%, SG as 9.7 seeds day^{-1}, and CG as 7.37% (Table 8.3).

8.3.3 Seedling Growth Parameters

Different parameters of seedling growth were evaluated in different populations after 21 days of seed germination. Among the different populations, MAA was found to be fast-growing seedlings as it showed significantly increased values of different parameters such as an average SH of 10.31 cm (64.3% over control), VI of 941.7 (198.2% over control), SFW 4.77 g (51.3% over control) per ten seedlings, and SDW of 0.90 g (187.2% over control) per ten seedlings after 21 days of seed germination (Table 8.4; Fig. 8.3). On the other hand, the least growing seedlings of MCD population showed minimum values of various parameters, that is, SH of 6.27 cm, VI of 315.8, SFW of 3.15 g per ten seedlings, and SDW of 0.31 g per ten seedlings, after 21 days of seed germination (Table 8.4).

8.3.4 Photosynthetic Pigment Contents

The comparative analysis of different photosynthetic pigments chlorophyll (*Chl* a/b, T*Chl*, and *Car*) was evaluated in 21-day-old seedlings of different population. The photosynthetic pigments content of MAA population showed a significantly higher concentration than seedlings of other populations (Table 8.5). Seedlings of MAA populations showed maximum values of *Chl* a 2.03 mg/g of FW (68.9% over control), *Chl* b 0.84 mg/g of FW (80.7% over control), T*Chl* 2.87 mg/g FW (73.3% over control), and *Car* 0.70 mg/g of FW (90.9% over control) whereas minimum values of *Chl* a 1.20, *Chl* b 0.47, T*Chl* 1.65, and *Car* 0.37 mg/g of FW were recorded in seedlings of MCD population (Table 8.5).

8.4 Discussion

Forest tree species are extremely important for the world economy as well as for maintaining and conservation of nature and natural resources. Many countries, including India, are facing acute problems in maintaining forest tree resources because of their uncontrolled exploitation through anthropogenic activities and more recently climate change. These activities have resulted in the sharp decline in the population of many plant species. According to International Union for Conservation of Nature (IUCN 2018), *P. marsupium* is an internationally threatened tree species as it is being continuously overexploited by several agencies and now has reached at the edge of extinction. The mass clonal propagation of elite species is necessary to meet growing demands for forest products and reforestation as well as application in tree improvement programs (Husain et al. 2010; Ahmed et al. 2017).

Table 8.3 Seed germination parameters evaluated after 21 days of seed culture of different populations of *P. marsupium*

	Populations	PG (%)		SG (seeds day^{-1})		CG (%)		MGT (days)	
S. N.	code	Mean ± SE	(PV)	Mean ± SE	(PV)	Mean ± SE	(PV)	Mean ± SE	(PV)
1	MMK	53.0 ± 0.58	$(5.3)^{i}$	11.0 ± 0.05	$(13.4)^{i}$	7.52 ± 0.02	$(1.9)^{i}$	13.30 ± 0.04	$(-1.9)^{b}$
2	MJH	62.0 ± 0.58	$(23.2)^{f}$	14.8 ± 0.03	$(52.8)^{f}$	7.76 ± 0.01	$(5.3)^{f}$	12.88 ± 0.02	$(-5.0)^{e}$
3	MAA	91.3 ± 0.33	$(81.5)^{a}$	29.0 ± 0.19	$(199.5)^{a}$	8.23 ± 0.01	$(11.6)^{a}$	12.15 ± 0.01	$(-10.4)^{j}$
4	MHP	65.0 ± 0.58	$(29.1)^{e}$	16.4 ± 0.16	$(69.0)^{e}$	7.83 ± 0.00	$(6.1)^{e}$	12.78 ± 0.00	$(-5.8)^{f}$
5	MBG	55.0 ± 0.58	$(9.3)^{h}$	12.1 ± 0.16	$(25.4)^{h}$	7.58 ± 0.01	$(2.8)^{h}$	13.20 ± 0.02	$(-2.7)^{c}$
6	MSS	72.3 ± 0.33	$(43.7)^{d}$	19.8 ± 0.11	$(104.5)^{d}$	7.97 ± 0.01	$(8.1)^{d}$	12.55 ± 0.01	$(-7.4)^{g}$
7	MCD (C)	50.3 ± 0.33	$(0.0)^{j}$	09.7 ± 0.05	$(0.0)^{j}$	7.37 ± 0.01	$(0.0)^{j}$	13.56 ± 0.03	$(0.0)^{a}$
8	MBS	78.0 ± 0.58	$(55.0)^{c}$	22.6 ± 0.11	$(133.4)^{c}$	8.06 ± 0.01	$(9.3)^{c}$	12.41 ± 0.01	$(-8.5)^{h}$
9	CSA	60.0 ± 0.58	$(19.2)^{g}$	13.5 ± 0.04	$(39.8)^{g}$	7.69 ± 0.02	$(4.2)^{g}$	13.01 ± 0.04	$(-4.1)^{d}$
10	CRK	81.0 ± 0.58	$(60.9)^{b}$	24.2 ± 0.04	$(149.4)^{b}$	8.10 ± 0.01	$(9.9)^{b}$	12.34 ± 0.02	$(-9.0)^{i}$

Values in parentheses are percentage variation (*PV*). Means ± SE values ($n = 3$) or three repeats of 100 seeds of each population followed by the same letter within column of each population are not significantly different ($p = 0.05$) using Duncan's multiple test. (*C*, Control; *PG*, percentage germination; *SG*, speed germination; *CG*, coefficient germination; *MGT*, mean germination time)

Table 8.4 Seedling growth parameters evaluated in different populations of *P. marsupium* after 21 days of seed germination

	Populations	SH (cm)		VI		SFW (g per 10 seedlings)		SDW (g per 10 seedlings)	
S. N.	code	Mean ± SE	(PV)	Mean ± SE	(PV)	Mean ± SE	(PV)	Mean ± SE	(PV)
1	MMK	06.52 ± 0.01	(3.9)^{g}	345.6 ± 3.62	(9.4)^{i}	3.19 ± 0.01	(1.4)^{h,i}	0.34 ± 0.02	(8.5)^{g}
2	MJH	06.90 ± 0.04	(10.0)^{f}	428.0 ± 2.33	(35.5)^{f}	3.40 ± 0.01	(8.0)^{f}	0.40 ± 0.01	(28.7)^{f}
3	MAA	10.31 ± 0.03	(64.3)^{a}	941.7 ± 6.00	(198.2)^{a}	4.77 ± 0.02	(51.3)^{a}	0.90 ± 0.02	(187.2)^{a}
4	MHP	07.28 ± 0.03	(16.0)^{e}	473.0 ± 6.05	(49.8)^{e}	3.52 ± 0.02	(11.9)^{e}	0.47 ± 0.00	(51.1)^{e}
5	MBG	06.58 ± 0.02	(4.9)^{g}	362.1 ± 2.99	(14.7)^{h}	3.24 ± 0.01	(2.9)^{h}	0.34 ± 0.02	(8.5)^{g}
6	MSS	07.68 ± 0.02	(22.5)^{d}	555.8 ± 3.26	(76.0)^{d}	3.76 ± 0.01	(19.4)^{d}	0.53 ± 0.00	(68.1)^{d}
7	MCD (C)	06.27 ± 0.01	(0.0)^{h}	315.8 ± 2.06	(0.0)^{j}	3.15 ± 0.01	(0.0)^{i}	0.31 ± 0.01	(0.0)^{g}
8	MBS	08.05 ± 0.04	(28.4)^{c}	628.2 ± 5.38	(98.9)^{c}	3.92 ± 0.05	(24.4)^{c}	0.64 ± 0.01	(103.2)^{c}
9	CSA	06.85 ± 0.02	(9.2)^{f}	411.0 ± 3.31	(30.2)^{g}	3.32 ± 0.02	(5.4)^{g}	0.40 ± 0.01	(26.6)^{f}
10	CRK	08.52 ± 0.01	(35.8)^{b}	690.1 ± 5.86	(118.6)^{b}	4.02 ± 0.02	(27.7)^{b}	0.67 ± 0.01	(113.8)^{b}

Values in parentheses are percentage variation (*PV*). Means ± SE values ($n = 3$) or three repeats of ten randomly selected seedlings of each population followed by the same letter within column of each population are not significantly different ($p = 0.05$) using Duncan's multiple test. (*C*, control; *SH*, seedling height; *VI*, vigor index; *SFW*, seedling fresh weight; *SDW*, seedling dry weight)

Table 8.5 Estimation of photosynthetic pigments in 21-day-old axenic seedlings of different populations of *P. marsupium*

	Populations	*Chl* a (mg/g of FW)		*Chl* b (mg/g of FW)		T*Chl* (a + b) (mg/g of FW)		*Car* (mg/g of FW)	
S. N.	code	Mean ± SE	(PV)	Mean ± SE	(PV)	Mean ± SE	(PV)	Mean ± SE	(PV)
1	MMK	1.29 ± 0.02	(7.8)^{h}	0.53 ± 0.03	(14.3)^{d,e}	1.80 ± 0.04	(8.7)^{g}	0.40 ± 0.01	(9.1)^{e,f}
2	MJH	1.58 ± 0.01	(31.4)^{e}	0.62 ± 0.04	(32.8)^{c,d}	2.18 ± 0.04	(31.9)^{e}	0.49 ± 0.02	(33.6)^{c,d}
3	MAA	2.03 ± 0.03	(68.9)^{a}	0.84 ± 0.01	(80.7)^{a}	2.87 ± 0.03	(73.6)^{a}	0.70 ± 0.01	(90.9)^{a}
4	MHP	1.69 ± 0.02	(40.8)^{d}	0.62 ± 0.02	(32.1)^{c,d}	2.31 ± 0.00	(39.5)^{d}	0.52 ± 0.04	(42.7)^{b,c,d}
5	MBG	1.36 ± 0.02	(13.3)^{g}	0.53 ± 0.03	(14.3)^{d,e}	1.88 ± 0.04	(13.9)^{g}	0.40 ± 0.01	(10.0)^{e,f}
6	MSS	1.82 ± 0.01	(51.9)^{c}	0.70 ± 0.01	(50.0)^{b,c}	2.52 ± 0.01	(52.6)^{c}	0.55 ± 0.02	(50.9)^{b,c}
7	MCD (C)	1.20 ± 0.02	(0.0)^{i}	0.47 ± 0.03	(0.0)^{e}	1.65 ± 0.02	(0.0)^{h}	0.37 ± 0.03	(0.0)^{f}
8	MBS	1.88 ± 0.02	(56.4)^{b,c}	0.72 ± 0.01	(55.0)^{b}	2.60 ± 0.04	(57.3)^{b,c}	0.58 ± 0.02	(57.3)^{b}
9	CSA	1.50 ± 0.01	(25.0)^{f}	0.57 ± 0.03	(21.4)^{d}	2.07 ± 0.03	(25.4)^{f}	0.45 ± 0.04	(22.7)^{d,e}
10	CRK	1.90 ± 0.02	(58.1)^{b}	0.74 ± 0.02	(58.6)^{b}	2.64 ± 0.04	(59.5)^{b}	0.59 ± 0.03	(61.8)^{b}

Values in parentheses are percentage variation (*PV*). Means values ($n = 3$) or three repeats of randomly selected seedlings of each population followed by the same letter within column of each population are not significantly different ($p = 0.05$) using Duncan's multiple test. (*C*, control; *Chl*, Chlorophyll; T*Chl*, total chlorophyll; *Car*, carotenoids)

Seed germination refers to the emergence of root or shoot from the seed coat, while emergence is the visible penetration of the shoot above the medium surface (Kader 2005). The seed germination experiment helps the researchers to determine if the seeds are suitable for planting. Studies regarding the seed germination are important for obtaining suitable planting materials for in vitro propagation (Ranal et al. 2009; Gairola et al. 2011; Singh and Kumar 2014; Trivedi and Joshi 2014). During in vitro germination practice, the optimization of an appropriate culture medium alone or in combination with appropriate plant growth regulators (PGRs) has provided useful information toward in vitro seed germination. In the present study, the ½ MS medium containing 0.5 μM GA_3 was found to be the most effective treatment and showed 81% of germination response in Raigarh (CRK) accessions. Augmentation of GA_3 in culture media has been reported in many plant species for improving germination along with better elongation of plantlets (Varshney et al. 2013; Hawkins 2018). It is also reported that GA_3 regulated the synthesis of α-amylase in aleuronic layer of seeds by upregulating α-amylase gene, SLN1 and GAMYB transcription factors promoting germination (Gubler et al. 2002), and DELLA-mediated inhibition of BZR1 transcription factor promotes plantlet elongation (Li and He 2013). The increased concentration of GA_3 favored to rupture the testa and endosperm. In many scientific reports, the plant species have developed a huge morphological and physiological diversity in seed types (Leugner et al. 2009; Deligoz 2012). Understanding the process of seed germination as a whole from early seed germination to the establishment of seedling can help plant breeders to choose better and healthier plant accessions (Ranal et al. 2009; Gairola et al. 2011).

In vitro germination and seedling growth parameters of our study showed that accessions of Amarkantak (MAA) were fast-growing compared to other populations. Analysis of germination can provide valuable information about the percentage rate, uniformity, and mean germination time (MGT) of seeds per day. By knowing the germination parameters, a researcher can adjust the planting rates to attain the desired plant population in the field. Thus, germination is considered to be a qualitative developmental response of populations, however, germination time varied among different accessions. The coefficient of germination gives an indication of the rapidity of germination of a population (Thomson and El-Kassaby 1993). It increased when number of germinated seed increases and time required for germination decreases (Jones and Sanders 1987; Oliveira and Garcia 2011). However, mean germination time is an accurate measure of time taken for a population seed to germinate, but does not correlate this well with the time spread or uniformity of germination (Matthews and Khajeh-Hosseini 2006, 2007). In the present study, MAA accession has attained minimum values of mean germination time while Chhindwara (MCD) accession showed maximum value of MGT. Our studies on germination, seedling height, biomass content, and vigor index of ten populations clearly correlated the growth pattern of all the accessions showing that MAA accession was fast-growing, while MCD accession showed slow growth in central Indian forest region. The study might help in the selection of superior mother plants prior to the development of an in vitro propagation protocol for producing healthy and disease-free plantlets that can be used for reforestation and sustainable consumption purposes.

8.5 Conclusion

To conclude, half-strength MS media in combination with GA_3 was found to be most effective on enhancement of maximum seed germination with an average seedling height. On the basis of seed germination parameters and seedling growth performance, Amarkantak (MAA) population was found to be significantly fast grown and Deogarh (MCD) was a slow grown population in central Indian forest region. The order of superiority pattern of ten populations on the basis of germination parameters and seedling growth performance was recorded as MAA > CRK > MBS > MSS > MHP > MJH > CSA > MBG > MMK > MCD in this study. To our knowledge, this is the first report to demonstrate the identification/selection of elite genotype of *P. marsupium* among the naturally grown ten populations via in vitro seed germination and seedling growth performance. The study may surely help in the selection of superior mother plants prior to develop an in vitro propagation protocol to be used in tree improvement program. It may also be helpful in producing healthy and disease-free plantlets that can be used for reforestation and sustainable consumption purposes.

Acknowledgments We acknowledge the Department of Biotechnology, Government of India, New Delhi for financial support (vide no. BT/PR2189/PBD/17/744/2011 Dated 19.12.2012). We are also thankful to the Forest Department of Madhya Pradesh and Chhattisgarh states of central India for providing planting materials.

References

Abdul-Baki A, Anderson JD (1973) Vigor determination in Soybean seed by multiple criteria. Crop Sci 13:630–633

Ahmad A, Ahmad N, Anis M (2018) Preconditioning of nodal explants in Thidiazuron-supplemented liquid media improves shoot multiplication in *Pterocarpus marsupium* (Roxb.). In: Ahmad N, Faisal M (eds) Thidiazuron: from urea derivative to plant growth regulator. Springer, Singapore, pp 175–185

Ahmed MR, Anis M, Alatar AA, Faisal M (2017) In vitro clonal propagation and evaluation of genetic fidelity using RAPD and ISSR marker in micropropagated plants of *Cassia alata* L.: a potential medicinal plant. Agrofor Syst 91:637–647

Anis M, Husain MK, Shahzad A (2005) In vitro plantlet regeneration of *Pterocarpus marsupium* Roxb., an endangered leguminous tree. Curr Sci 88:861–863

Barstow M (2017) *Pterocarpus marsupium*. The IUCN Red List of Threatened Species 2017

Chakraborty A, Gupta N, Ghosh K, Roy P (2010) In vitro evaluation of the cytotoxic, anti-proliferative and anti-oxidant properties of pterostilbene isolated from Pterocarpus marsupium. Toxicol in Vitro 24:1215–1228

Chand S, Singh AK (2004) In vitro shoot regeneration from cotyledonary node explants of a multipurpose leguminous tree, *Pterocarpus marsupium* Roxb. In Vitro Cell Dev Biol Plant 40:464–466

Copeland LO (1976) Principals of seed science and technology. Burgess Publishing Company, Minnaeapolis, Minnesofa, pp 164–165

Deligoz A (2012) Morphological and physiological differences between bareroot and container *Juniperus excelsa* seedlings. Turk J Agricult Forest 36(5):619–628

Dhanabal SP, Kokate CK, Ramanathan M, Kumar EP, Suresh B (2006) Hypoglycaemic activity of *Pterocarpus marsupium* Roxb. Phytother Res 20:4–8. e.T34620A67802995. https://doi.org/10.2305/IUCN.UK.20173.RLTS.T34620A67802995.en

Ellis RH, Roberts EH (1981) The quantification of ageing and survival in orthodox seeds. Seed Sci Technol 9:377–409

Gairola KC, Nautiyal AR, Dwivedi AK (2011) Effect of temperatures and germination media on seed germination of *Jatropha curcas* Linn. Adv Biores 2:66–71

Gamborg OL, Miller R, Ojima K (1968) Nutrient requirements of suspension cultures of soybean root cells. Exp Cell Res 50:151–158

Gubler F, Chandler PM, White RG, Llewellyn DJ, Jacobsen JV (2002) Gibberellin signaling in barley aleurone cells. Control of SLN1 and GAMYB expression. Plant Physiol 129:191–200

Hawkins TS (2018) Regulating acorn germination and seedling emergence in *Quercus pagoda* (Raf.) as it relates to natural and artificial regeneration. New For 50:425–536

Hougee S, Faber J, Sanders A, de Jong RB, van den Berg WB, Garssen J, Hoijer MA, Smit HF (2005) Selective COX-2 inhibition by a *Pterocarpus marsupium* extract characterized by pterostilbene, and its activity in healthy human volunteers. Planta Med 71:387–392

Husain MK, Anis M, Shahzad A (2010) Somatic embryogenesis and plant regeneration in *Pterocarpus marsupium* Roxb. Tree Struct Funct 24:781–787

ISTA (1999) International rules for seed testing. Seed Sci Technol 21:288

IUCN (2018) *Pterocarpus marsupium*. doi:https://doi.org/10.2305/IUCN.UK.2017-3.RLTS.T34620A67802995.en. Accessed 25 June 2018

Jones K, Sanders D (1987) The influence of soaking pepper seed in water or potassium salt solutions on germination at three temperatures. J Seed Technol 11:97–102

Kader MA (2005) A comparison of seed germination calculation formulae and the associated interpretation of resulting data. J Proc R Soc New South Wales 138:65–75

Leugner J, Jurásek A, Martincová J (2009) Comparison of morphological and physiological parameters of the planting material of Norway spruce (*Picea abies* [L.] Karst.) from intensive nursery technologies with current bare root plants. J Forest Sci 55:511–517

Li QF, He JX (2013) Mechanism of signalling crosstalk between brassinosteroids and gibberelins. Plant Signal Behav 8:e24686

Lloyd G, McCown B (1980) Commercially feasible micropropagation of mountain laurel, Kalmia latifolia, by use of shoot-tip culture. Proc Int Plant Propagat Soc 30:421–427

Mackinney G (1941) Absorption of light by chlorophyll solution. J Biol Chem 140:315–322

Maguire JD (1962) Speed of germination- aid in selection and evaluation for seedling emergence and vigour. Crop Sci 2:176–177

Matthews S, Khajeh-Hosseini M (2006) Mean germination time as an indicator of emergence performance in soil of seed lots of maize (*Zea mays*). Seed Sci Technol 34:339–347

Matthews S, Khajeh-Hosseini M (2007) Length of the lag period of germination and metabolic repair explain vigour differences in seed lots of maize (*Zea mays*). Seed Sci Technol 35:200–212

Maurya R, Singh R, Deepak M, Handa SS, Yadav PP, Mishra PK (2004) Constituents of *Pterocarpus marsupium*: an ayurvedic crude drug. Phytochemistry 65:915–920

Mohire NC, Salunkhe VR, Bhise SB, Yadav AV (2007) Cardiotonic activity of aqueous extract of heartwood of *Pterocarpus marsupium*. Indian J Exp Biol 45:532–537

Murashige T, Skoog F (1962) A revised medium for rapid growth bioassays with tobacco tissue culture. Physiol Plant 15:473–497

NMPB (2018). http://nmpb.nic.in/medicinal_list. Accessed 25 June 2018

Oliveira PG, Garcia QS (2011) Germination characteristics of *Syngonanthus* seeds (*Eriocaulaceae*) in *campos rupestres* vegetation in south-eastern Brazil. Seed Sci Res 21:39–45

Ranal MA, Santana DGD, Ferreira WR, Rodrigues CM (2009) Calculating germination measurements and organizing spreadsheets. Braz J Bot 32:849–855

Remsberg CM, Yáñez JA, Ohgami Y, Vega-Villa KR, Rimando AM, Davies NM (2008) Pharmacometrics of pterostilbene: preclinical pharmacokinetics and metabolism, anticancer, anti-inflammatory, antioxidant and analgesic activity. Phytother Res 22:169–179

Sanders A, Smit HF, Garssen J, Faber J, Hoijer MA (2005) *Pterocarpus marsupium* extract exhibits anti-inflammatory activity in human subjects. Plant Med 71:387–392

Singh VP, Kumar D (2014) Influence of seed size and pretreatments on seed germination and seedling vigour of *Pterocarpus marsupium* Roxb. Ind Forest 140:887–890

Thomson AJ, El-Kassaby YA (1993) Interpretation of seed germination parameters. New For 7:123–132

Tippani R, Porika M, Allenki V, Anreddy RNR, Yellu NR, Krishna DR, Thammidala C, Abbagani S (2010) Antioxidant and analgesic activities of *Pterocarpus marsupium* Roxb. J Herbs Spices Med Plant 16:63–68

Trivedi DR, Joshi AG (2014) Studies on seed germination of *Stereospermum suaveolens* with respect to different parameters. Environ Exp Biol 12:33–37

Varshney A, Anis M, Aref IM (2013) Control of bioregulants on plant resurgence in vitro from mature seeds of Egyptian myrobalan tree (*Balanites aegyptiaca* Del.)–A plant affluent in saponins. Int J Pharmaceut Sci Rev Res 22:23–28

Regulation of In Vitro Morphogenesis by Modulation of Culture Conditions in *Withania somnifera* L. Using Cotyledonary Node Explants

9

Nigar Fatima and M. Anis

Abstract

Optimization of culture conditions has been proposed for high-frequency shoots regeneration by in vitro propagation of cotyledonary node segments (CN) attained from 7-day-old aseptic seedlings in a highly valuable medicinal plant—*Withania somnifera* L. Micropropagation was accomplished from CN explants as bared to diverse range of cytokinins viz; 6-benzyl adenine (BA), kinetin (Kn), or 2-isopentenyl adenine (2iP) individually or in combined arrangement with various auxins namely indole 3-acetic acid (IAA), indole-3-butyric acid, or α-naphthalene acetic acid (NAA) in Murashige and Skoog medium. Bud break was dependent on cytokinin supply and the maximum rate (95%) of regeneration with highest number of shoots (13.0 ± 0.00) and shoot length (2.00 ± 0.36 cm) was evident on MS medium composed of BA (2.5 μM), after 4 weeks of incubation. However, the synergistic combination of cytokinin, BA (2.5 μM) along with auxin NAA (0.5 μM) encouraged maximum number (29.3 ± 0.23) of shoots and shoot length (5.62 ± 0.17 cm) with 90% regeneration frequency after 8 weeks of culture. After 4 weeks, the greatest number 12.6 ± 0.10 of roots with root length of 3.58 ± 0.16 cm/shoot with 95% regeneration frequency were attained on medium fortified with NAA (0.5 μM). The results highlight the potential conservation approaches for constant furnish of plant materials for both local and escalating markets as well as safety evaluation on sustainable usage of the species for medicinal purposes.

Keywords

Acclimatization · Cotyledonary node explants · Micropropagation · Morphogenesis · Plant growth regulators · *Withania somnifera*

N. Fatima (✉) · M. Anis
Plant Biotechnology Laboratory, Department of Botany, Aligarh Muslim University, Aligarh, India

I. Siddique (ed.), *Propagation and Genetic Manipulation of Plants*,
https://doi.org/10.1007/978-981-15-7736-9_9

Abbreviations

2-iP	2-Isopentenyl adenine
BA	6-Benzyladenine
CN	Cotyledonary node
IBA	Indole-3-butyric acid
Kin	6-Furfurylaminopurine
MS	Murashige and Skoog's medium
NAA	α-Naphthalene acetic acid

9.1 Introduction

Ashwagandha (*Withania somnifera* L.), a recognized medicinal plant sharing numerous health prosperity and well-being, a vital "Rasayana" marketed as "Sattvic Kapha Rasayana" (Indian ayurveda) since centuries designed for its incredible benefits (Mahima et al. 2012). A diverse spectrum of phytochemicals present in plant facilitates it to hold an extensive assortment of biological implications. Roots, stems, leaves, and fruits/berries of plant show unique medicinal properties approved to their assorted metabolites (Bharti et al. 2011; Bhatia et al. 2013). Extensive range of therapeutic activities is corroborated by the existence of steroidal lactones present in plant roots, which promote and maintain general health. Immunomodulation, anticancer and antiepileptic, memory enhancer, mood elevator, diuretic, rejuvenator, stress reliever, antiaging, antioxidant, hypoglycemic, adaptogen, and so on, are major health benefits and veterinary sciences focus on its potent function in maintaining sound health and therapeutic usages (Tiwari et al. 2014). Ashwagandha's extensive pharmacological activities prepared the plants as an essential constituent of polyherbal preparations. A number of central preparations like Asvagandharishtam, chavanaprasam, Valiya Narayana tailam, and so on, using the drugs are available in the market. However, conventional cultivation via seed is not sufficient to hold the flourishing requirement of *Withania* for medicament needs.

Withania propagates via seeds but low seed germination percentage, seed viability period, and shrinkage of innate potentials for vegetative propagation (Sen and Sharma 1991; Farooqi and Sreeramu 2004), resulted in deterioration of plant materials, making the plant at a risk of extinction in wild, hindrance in plant cultivation and depletion in market values. Pharmaceutical corporations are unearthing this medicinal herb from the nature, creating a huge pressure to its existence as well as troubling environmental stability. This plant species is also incorporated in the list of endangered species by International Union for Conservation of Natural Resources (Supe et al. 2006; Patel and Krishnamurthy 2013).

Modern plant improvement programs (biotechnological inventions) include in vitro tissue culture practices designed for aseptic plants production, introduction of novel traits into preferred plants, multiplication of elite selections, and development of appropriate cultivars in the small duration (Anis and Ahmad 2016). In vitro manipulations and propagation of plants from juvenile seedlings containing cotyledonary node (CN) as an explant have earned significant attentions currently because of its simple accessibility, speedy response, and immense potential in favor of shoot organogenesis, somatic embryogenesis, culture of protoplast, and transformation comprehension (Tabei et al. 1991; Compton et al. 1994).

Phytohormones (auxins and cytokinins) recognized as the key signal molecules in the development and growth of plants (Moubayidin et al. 2009). Cytokinins activate the release of dormant shoot sprouts. In general, different cytokinins articulate different events in manipulating in vitro axillary shoot formation (Preece and Shutter 1991). In vitro cultures of plant cells, tissues, and organs are not exclusively autotrophic; ascertain requirements for continuous supply of energy sources (carbohydrate) in culture media which acts as energy sources and also maintains osmotic potential (Lipavaska and Konradova 2004). Moreover, various developmental processes like shoot proliferation, root induction, organogenesis, embryogenesis are highly energy-requiring procedures in plant biology, occur at the expense of carbon sources (De Klerk and Calamar 2002). pH of medium is an essential factor of physiochemical background throughout the progress of plant tissues, organs in vitro (Williams et al. 1990). Optimum pH is imperative for improved plant growth and expansion, the suboptimal pH levels leading to abnormalities during explants development (Anderson and Levinsh 2008).

Till date, only a single report is available for *W. somnifera* propagation and regeneration in vitro utilizing CN explants (Nayak et al. 2012). However, the report did not provide comprehensive studies on the collective outcomes of plant growth regulators, media types, carbon supply, and pH factors on plants regeneration system. Therefore, contemplating the need and relevance of the plant, the current experiment was designed to formulate a refined, resourceful, and reproducible regeneration system in Ashwagandha by optimizing various parameters like plant growth regulators (PGRs), media types, carbon sources, and medium pH.

9.2 Materials and Methods

9.2.1 Seed Germination and Preparation of Explants

Seeds of *W. somnifera* were procured from superior genotypes established at the Botanical Garden of the Aligarh Muslim University, Aligarh. Thorough washing of seeds was carried out for 30 min under running tap water in order to eliminate adherent particles. The seeds were treated with [5% (v/v)] freshly prepared Labolene, a liquid detergent (Qualigens, Mumbai, India), for 20 min and finally rinsed with distilled water (sterile DW). The surface sterilization was undertaken with [0.1% (w/v)] mercuric chloride ($HgCl_2$) solution (Qualigens, India) for 5 min and lastly washed five times with sterile distilled water for the removal of sterilant traces before implantation. The sterilized seeds were aseptically germinated on different strength (full, 1/2, 1/3, and 1/4) of Murashige and Skoog (1962) medium to raise aseptic seedlings. Cotyledonary nodes (7 day old) were used as explants in the experiments.

9.2.2 Culture Media and Conditions

In the experiments, Murashige and Skoog (MS) medium fortified with sucrose 3% (w/v) and a bacteriological grade agar (Hi-media) 0.8% (w/v) was employed. 5.8 medium pH was adjusted by 1 N NaOH/1 N HCl prior to autoclaving the culture media at pressure (121 °C at 1.06 kg cm^{-2}) for 20 min. The entire cultures were kept in culture room with standard environmental conditions of temperature, humidity, and photoperiod duration, which includes photoperiod (16 h), photosynthetic photon flux density (PPFD) (50 μmol m^{-2} s^{-1}), 40 W fluorescent tubes (Phillips, India), 50–60% relative humidity under 24 ± 2 °C.

9.2.3 Induction and Multiplication Shoot

Seven-day-old aseptic CN explants containing axillary buds were placed on different concentrations (1.0, 2.5, 5.0, 7.5, and 10.0 μM) of various cytokinins (BA, Kn, and 2iP) alone or else in mixture with diverse auxins (IAA, IBA, and NAA) at a range of concentrations (0.5, 1.0, and 2.0 μM) in MS medium for shoot bud initiation and regeneration. Multiplication of shoot and stable/constant establishment, the regenerating tissues were recurrently subcultured on the regenerative medium [MS + BA (2.5 μM) + NAA (0.5 μM)] after every 6 weeks.

9.2.4 Effect of Different Media, pH, and Carbon Sources

CN segments were cultured on diverse basal media specifically MS (Murashige and Skoog 1962), B_5 medium (Gamborg et al. 1968), L_2 (Phillips and Collins 1979), and WPM (Lloyd and McCown 1981), carbon sources (glucose, sucrose, and fructose) along with media pH (4.2, 5.2, 6.4, and 7.0) containing most favorable concentration of BA (2.5 μM) plus NAA (0.5 μM) for greatest shoot regeneration and proliferation. Subculturing of cultures was carried out onto fresh media at 3 weeks interval. Explants frequency, quantity of shoot/explants with shoot length after 8 weeks of culture conditions were recorded regularly.

9.2.5 Rooting and Acclimatization

In vitro rooting was carried out through the excision and transfer of (3–4 cm) long microshoots on rooting medium containing various auxins applications (0.1–2.0 μM) viz., indole-3-acetic (IAA), indole-3-butyric acid (IBA), and α-naphthalene acetic acid (NAA). The rooted plantlets were potted in plastic pots containing sterile soilrite™ (Keltech, India Ltd., Bangalore). The potted plantlets were covered with transparent polythene bags to guarantee elevated humidity, also watered with half-strength MS salts upto 2 weeks, and placed below 16/8 h light photoperiod environment. Toward the process of plantlets acclimatization and hardening, the polythene bags were opened frequently. Further, following 4 weeks period, the adapted plants were shifted to earthen pots holding regular garden soil, sustained in a net house under innate light. Rooting percentage, root numbers with root length data were documented prior to relocating into soil.

9.2.6 Data Collection and Statistical Analysis

A minimum of 20 replicates/treatments with three times repetition of experiment were carried out. The periodical and morphological modification in the cultures was observed and documented at regular period. Statistical analysis of the data was prepared employing SPSS version 16 (SPSS Inc., Chicago, USA) and the significance of differences was conceded out using Duncan's multiple range test (DMRT) at $P \leq 0.05$. The outcome was articulated as: means $\pm$ SE for all the experiments.

9.3 Results and Discussion

The implementation of plant cell, tissue, and organs in culture with significant application in the regulation of morphogenesis in plants is highly dependent on the utilization of appropriate PGR regimes, explants type, different factors like media types, pH, sucrose, light, temperature, and so on (Rout et al. 2000; Ahmad and Anis 2010; Fatima et al. 2015). Explants establishment is a very crucial stage for in vitro propagation and development. Plant growth regulators (PGRs) include a large arrangement of molecules that manipulate physiological and biological processes of higher plants at low concentrations (George et al. 2008). In general, these compounds have a regulatory role in plant growth, differentiation, and development. In vitro morphogenetic outcomes are largely governed by the presence and relative abundance of PGRs supplemented in culture media through complex pathways interacting with one another via multiple levels of control and regulation.

9.3.1 Establishment of Aseptic Seedlings

The surface-sterilized seeds were inoculated on different strengths (full, 1/2, 1/3, and 1/4) of MS medium for germination. Half-strength MS medium was found to be superior over other strengths. Maximum percentage (86.2%) of germination was recorded on ½ strength Murashige and Skoog medium, after 4 weeks of incubation (Fig. 9.1). On reducing the strength to 1/3 and 1/4, the germination response was found to be declined (Fig. 9.1). Seven-day-old CN excised from aseptic sapling was utilized as explants material.

Fig. 9.1 Effect of different strength of MS medium on in vitro germination of *W. somnifera* seeds, after 4 weeks of culture. Bars represent means ± SE. Bars denoted by the same letter within response variables are not significantly different ($P = 0.05$) using Duncan's multiple range test

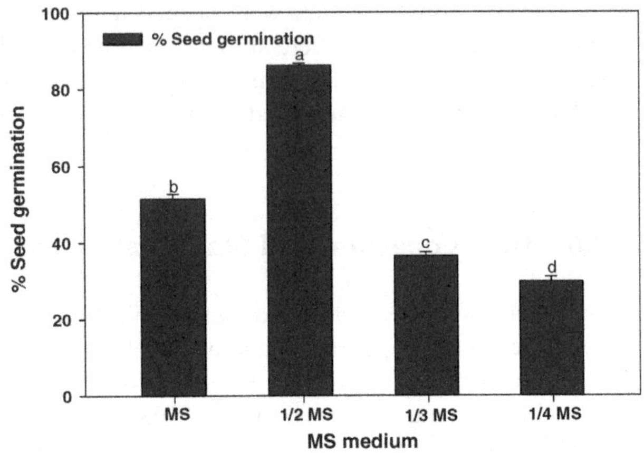

9.3.2 Implication of PGRs on Shoot Regeneration Outcome

CN explants did not respond morphogenetically to MS medium lacking growth regulator but once cultivated on medium complementing with different cytokinins namely BA, Kin, and 2iP (0.0–10.0 μM), a discrepancy in response was monitored (Table 9.1). The cotyledonary node induction is proved to be an efficient micropropagation method in plant as the emerging buds commencing from meristematic organs and tissues possess an immense potential for dynamic development (Yadav et al. 1990). Maximum rate of shoot sprout initiation was attained on Murashige and Skoog medium fortified with BA (2.5 μM), which gave maximum shoots (13.0 ± 0.00), (18.3 ± 0.20) with shoot length (2.00 ± 0.36 cm), (4.20 ± 0.12 cm) respectively, following 4 and 8 weeks of culture conditions (Table 9.1; Fig. 9.5a). For shoot stimulation and propagation in plant tissue culture system of numerous medicinal herbs, the superiority of BA over other cytokinins has been well discussed (Fatima et al. 2012; Faisal et al. 2018). A negative outcome with decline in regeneration frequencies and shoot bud formation from each explant is exhibited with an increase in BA concentration beyond the optimal level (2.5 μM). Cytokinin has been accounted for stimulating the activities and initiation of axillary meristems in the direction of shoot development. Medium having 2.5 μM Kin stimulated maximum shoot number (11.0 ± 0.00; 14.2 ± 0.20) with shoot length (1.36 ± 0.00; 3.50 ± 0.00 cm), after 4 and 8 weeks, respectively (Table 9.1). Whereas, 2iP augmented medium upshot in highest figure of shoots (10.8 ± 0.30; 12.4 ± 0.40) with shoot length (2.40 ± 0.22; 4.00 ± 0.00 cm) after 4 and 8 weeks of culture, respectively (Table 9.1). On elevated concentration of cytokinin, reduced shoot number and shoot length were examined. Whereas, on lowering the cytokinin application, stunted and malformed shoots were observed. The results obtained are in consonance with earlier findings in plants like *W. somnifera* (Fatima and Anis 2012) and *Cuphea procumbens* (Fatima et al. 2012), and so on. The decline in the regeneration prospective emerged owing toward a negative result of elevated applications on the cells programmed to outline vegetative buds.

Table 9.1 Effect of different cytokinins in MS medium on shoot regeneration from cotyledonary node explants in *W. somnifera* after 4 and 8 weeks of culture

Cytokinins (μM)				4 weeks		8 weeks	
BA	Kin	2iP	Regeneration (%)	Mean number of shoots	Mean shoot length (cm)	Mean number of shoots	Mean shoot length (cm)
0.0	–	–	00	0.00 ± 0.00^{g}	0.00 ± 0.00^{g}	0.00 ± 0.00^{o}	0.00 ± 0.00^{f}
0.1	–	–	70	5.04 ± 0.24^{f}	$0.50 \pm 0.10^{g,f}$	4.00 ± 0.00^{l}	1.10 ± 0.10^{e}
0.5	–	–	75	6.56 ± 0.36^{e}	$0.66 \pm 0.00^{d,e,f}$	$6.40 \pm 0.10^{i,j}$	1.36 ± 0.15^{e}
1.0	–	–	80	$10.40 \pm 1.37^{b,c}$	$1.20 \pm 0.00^{a,b,c,d}$	13.5 ± 0.16^{b}	$3.10 \pm 0.10^{b,c}$
2.5	–	–	95	13.00 ± 0.00^{a}	2.00 ± 0.36^{a}	18.3 ± 0.20^{a}	4.20 ± 0.12^{a}
5.0	–	–	70	10.90 ± 0.53^{b}	$2.56 \pm 0.14^{a,b,c,d}$	11.3 ± 0.22^{d}	2.80 ± 0.25^{c}
10.0	–	–	70	7.00 ± 0.00^{e}	$1.94 \pm 0.00^{c,d,e,f}$	7.20 ± 0.20^{h}	2.10 ± 0.10^{d}
	0.1	–	55	4.08 ± 0.23^{f}	$1.50 \pm 0.15^{e,f}$	3.34 ± 0.18^{m}	1.36 ± 0.41^{e}
	0.5	–	80	4.80 ± 0.30^{f}	$0.88 \pm 0.42^{c,d,e,f}$	5.10 ± 0.10^{k}	1.50 ± 0.00^{e}
	1.0	–	80	9.00 ± 0.31^{d}	$1.32 \pm 0.16^{c,d,e,f}$	10.4 ± 0.18^{e}	2.00 ± 0.00^{d}
	2.5	–	92	11.00 ± 0.00^{b}	$1.36 \pm 0.00^{a,b,c,d}$	14.2 ± 0.20^{b}	3.50 ± 0.00^{b}
	5.0	–	80	$9.48 \pm 0.38^{c,d}$	$2.00 \pm 0.22^{a,b,c}$	8.44 ± 0.19^{g}	2.14 ± 0.21^{d}
	10.0	–	80	7.48 ± 0.12^{e}	$2.10 \pm 0.00^{c,d,e,f}$	6.00 ± 0.00^{j}	1.20 ± 0.12^{e}
		0.1	70	4.00 ± 0.00^{f}	$1.50 \pm 0.27^{e,f}$	2.36 ± 0.15^{n}	1.20 ± 0.13^{e}
		0.5	75	5.00 ± 0.00^{f}	$1.00 \pm 0.36^{b,c,d,e}$	4.06 ± 0.11^{l}	1.32 ± 0.19^{e}
		1.0	82	$9.76 \pm 0.56^{b,c,d}$	$1.56 \pm 0.40^{a,b}$	$6.80 \pm 0.20^{h,i}$	$3.12 \pm 0.22^{b,c}$
		2.5	90	10.80 ± 0.30^{b}	2.40 ± 0.22^{a}	12.4 ± 0.40^{d}	4.00 ± 0.00^{a}
		5.0	80	7.00 ± 0.00^{e}	2.62 ± 0.60^{a}	6.02 ± 0.12^{j}	3.30 ± 0.12^{b}
		10.0	70	4.00 ± 0.00^{f}	$2.60 \pm 0.38^{c,d,e,f}$	4.00 ± 0.00^{l}	2.20 ± 0.10^{d}

Values represent means ± standard error of 20 replicates per treatment in three repeated experiments. Means followed by the same letter are not significantly different ($P = 0.05$) using Duncan's multiple range test

Table 9.2 Effect of optimal concentration of BA along with auxins in MS medium on shoot proliferation from cotyledonary node explants after 8 weeks of culture

Plant growth regulators (μM)				Regeneration (%)	Mean number of shoot	Mean shoot length (cm)
BA	IAA	IBA	NAA			
2.5	0.1	–	–	60	7.74 ± 0.18^{k}	3.60 ± 0.10^{d}
2.5	0.5	–	–	72	16.1 ± 0.16^{e}	$5.20 \pm 0.20^{a,b}$
2.5	1.0	–	–	70	11.4 ± 0.18^{g}	$4.12 \pm 0.22^{c,d}$
2.5	2.0	–	–	68	8.50 ± 0.15^{j}	3.00 ± 0.00^{e}
2.5	–	0.1	–	70	10.5 ± 0.16^{h}	4.36 ± 0.16^{c}
2.5	–	0.5	–	80	18.3 ± 0.16^{d}	5.00 ± 0.00^{b}
2.5	–	1.0	–	70	14.6 ± 0.18^{f}	2.50 ± 0.00^{e}
2.5	–	2.0	–	70	9.72 ± 0.18^{i}	$4.04 \pm 0.24^{c,d}$
2.5	–	–	0.1	80	22.4 ± 0.18^{b}	$4.00 \pm 0.00^{c,d}$
2.5	–	–	0.5	90	29.3 ± 0.23^{a}	5.62 ± 0.17^{a}
2.5	–	–	1.0	72	20.2 ± 0.15^{c}	4.24 ± 0.26^{c}
2.5	–	–	2.0	70	14.2 ± 0.20^{f}	3.56 ± 0.36^{d}

Values represent means ± standard error of 20 replicates per treatment in three repeated experiments. Means followed by the same letter are not significantly different ($P = 0.05$) using Duncan's multiple range test

Consequently, one of the most influencing factors in multiple shoot regeneration determining the modulation of endogenous auxin to cytokinin stability. Auxins and cytokinins have been extensively reviewed with regard to their biosynthesis, catabolism, transport, physiological roles, gene expression, and signal cascade (Müller and Leyser 2011; Shabir et al. 2016). Moreover, cytokinin–auxin ratio in culture medium plays an important factor determining the pathway of morphogenesis in plants in vitro. Ninety percent regeneration frequency with highest quantity of shoots (29.3 ± 0.23)/explant and shoot length (5.62 ± 0.17 cm) was observed on MS medium containing BA (2.5 μM) along with NAA (0.5 μM) followed by 8 weeks (Table 9.2; Fig. 9.5b). MS medium supplemented with Kin (2.5 μM) along with NAA (0.5 μM) induced maximum number of shoots (25.6 ± 0.17) with shoot length (5.26 ± 0.18 cm) per explant by 8 week (Table 9.3). Whereas, 2ip (2.5 μM) in combination with NAA (0.5 μM) proved lesser numbers (20.8 ± 0.30) of shoots and shoot length (5.88 ± 0.22 cm) per explant in the same duration (Table 9.4).

Escalating NAA concentration to 1.0 μM directed a decline in shoot arrangement competency along with basal callusing. Though, a gradual reduction in shoot numbers was examined at elevated concentrations of NAA (1.0 and 2.0 μM). NAA distinctly improved the regeneration percent, number of shoots, shoot lengths, while IBA or IAA did not progress the parameters estimated. Among different auxins analyzed, NAA was established to be most efficient followed by IBA and IAA in development of shoot clusters. The promising arrangement of BA along with NAA for in vitro shoot stimulation and multiplication has been accounted in a variety of medicinal plants such as *Ziziphora tenuiror* L. (Dakah et al. 2014), *Cassia*

Table 9.3 Effect of optimal concentration of Kin along with auxins in MS medium on shoot proliferation from cotyledonary node explants after 8 weeks of culture

Plant growth regulators (μM)				Regeneration (%)	Mean number of shoot	Mean shoot length (cm)
Kin	IAA	IBA	NAA			
2.5	0.1	–	–	72	15.1 ± 0.22^{f}	$4.36 \pm 0.16^{c,d}$
2.5	0.5	–	–	80	17.5 ± 0.36^{d}	$5.00 \pm 0.00^{a,b}$
2.5	1.0	–	–	75	14.0 ± 0.00^{g}	$5.08 \pm 0.48^{a,b}$
2.5	2.0	–	–	70	12.0 ± 0.00^{h}	$4.00 \pm 0.00^{c,d}$
2.5	–	0.1	–	80	16.4 ± 0.40^{e}	$4.12 \pm 0.22^{c,d}$
2.5	–	0.5	–	85	18.1 ± 0.22^{d}	5.20 ± 0.20^{a}
2.5	–	1.0	–	75	10.0 ± 0.63^{i}	$4.72 \pm 0.32^{a,b,c}$
2.5	–	2.0	–	65	7.60 ± 0.10^{j}	3.56 ± 0.36^{d}
2.5	–	–	0.1	80	24.3 ± 0.30^{b}	$4.00 \pm 0.00^{c,d}$
2.5	–	–	0.5	95	25.6 ± 0.17^{a}	5.26 ± 0.18^{a}
2.5	–	–	1.0	85	20.0 ± 0.48^{c}	$5.04 \pm 0.24^{a,b}$
2.5	–	–	2.0	70	15.0 ± 0.00^{f}	$4.08 \pm 0.23^{c,d}$

Values represent means ± standard error of 20 replicates per treatment in three repeated experiments. Means followed by the same letter are not significantly different ($P = 0.05$) using Duncan's multiple range test

Table 9.4 Effect of optimal concentration of 2iP along with auxins in MS medium on shoot proliferation from cotyledonary node explants after 8 weeks of culture

Plant growth regulators (μM)				Regeneration (%)	Mean number of shoot	Mean shoot length (cm)
2ip	IAA	IBA	NAA			
2.5	0.1	–	–	62	10.0 ± 0.00^{f}	3.28 ± 0.18^{g}
2.5	0.5	–	–	78	$14.0 \pm 0.00^{d,e}$	$5.00 \pm 0.00^{c,d}$
2.5	1.0	–	–	70	13.1 ± 0.74^{e}	$4.88 \pm 0.28^{c,d,e}$
2.5	2.0	–	–	70	8.40 ± 0.40^{g}	$4.30 \pm 0.30^{e,f}$
2.5	–	0.1	–	80	13.8 ± 0.28^{e}	$4.80 \pm 0.30^{c,d,e}$
2.5	–	0.5	–	85	16.0 ± 0.00^{c}	$5.68 \pm 0.17^{a,b}$
2.5	–	1.0	–	75	15.2 ± 0.70^{c}	4.00 ± 0.00^{f}
2.5	–	2.0	–	70	12.9 ± 0.31^{e}	$4.50 \pm 0.00^{d,e,f}$
2.5	–	–	0.1	78	18.0 ± 0.00^{b}	$5.14 \pm 0.21^{b,c,d}$
2.5	–	–	0.5	90	20.8 ± 0.30^{a}	5.88 ± 0.22^{a}
2.5	–	–	1.0	80	17.5 ± 0.36^{b}	$5.38 \pm 0.23^{a,b,c}$
2.5	–	–	2.0	75	$15.0 \pm 0.00^{c,d}$	$5.04 \pm 0.24^{b,c,d}$

Values represent means ± standard error of 20 replicates per treatment in three repeated experiments. Means followed by the same letter are not significantly different ($P = 0.05$) using Duncan's multiple range test

angustifolia (Siddique et al. 2015) and *Ruta graveolens* (Faisal et al. 2018). Wickson (1958) previously elucidated the promising regulatory cytokinin–auxin actions with supporting functions in apical dominance in shoot rejuvenation practice as well.

9.3.3 Effect of Media Types on Morphogenesis

MS medium was set up en route for the most appropriate medium designed for initiation of highest shoot buds as well as plant regeneration than B_5, L_2 and WPM. MS medium was established as more decisive when evaluated with other media for subsequent regeneration and proliferation of shoots (Fatima et al. 2015). WPM displayed deprived response for every parameters assessed and proved lesser number of shoots (11.2 ± 0.20) per explant, while L_2 gave agreeable results as compared to MS medium with a utmost of (15.0 ± 0.00) shoots/explant (Fig. 9.2). The highest number (20.0 ± 0.00) of shoots was recorded on full strength MS medium followed by L_2, B5, and WPM media in decreasing order when observed after 4 weeks of culture (Fig. 9.2).

9.3.4 Effect of Diverse Carbon Sources

Different applications of carbon sources were added in the culture medium to evaluate their effects on shoot regeneration and growth in *W. somnifera*. However, differential responses were recorded in plant morphogenesis with varied concentrations of carbon sources. It is well established that the carbohydrate requirements depend on the culture stages and might prove variation in uptake and utilization of different carbon sources with different species. The effect of different carbon sources (sucrose, fructose, and glucose) with a most approving medium holding BA (2.5 μM) along with NAA (0.5 μM) was studied for shoot regeneration.

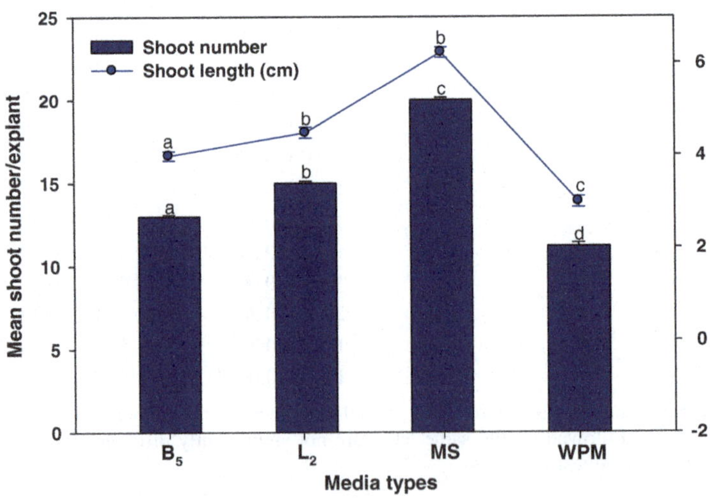

Fig. 9.2 Effect of different media types on shoot proliferation from cotyledonary node explants supplemented with BA (2.5 μM) and NAA (0.5 μM) after 4 weeks of culture. Bars represent the means ± SE. Bars denoted by the same letter within response variables are not significantly different ($P = 0.05$) using Duncan's multiple range test

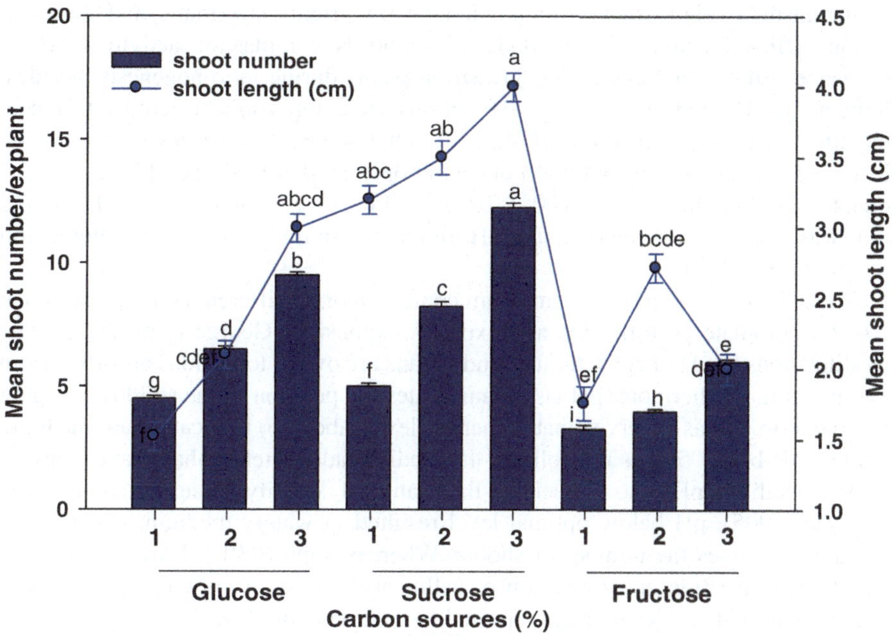

Fig. 9.3 Effect of different carbon sources on shoot proliferation from cotyledonary node explants supplemented with BA (2.5 μM) and NAA (0.5 μM) after 4 weeks of culture. Bars represent means ± SE. Bars denoted by the same letter within response variables are not significantly different ($P = 0.05$) using Duncan's multiple range test

Three percent sucrose (w/v) induced an optimum number of shoots (23.4 ± 0.40) followed by glucose (12.6 ± 0.10) and fructose (8.20 ± 0.27), respectively (Fig. 9.3).

Carbohydrate controls in vitro morphogenesis by functioning as energy resource, adjusting the osmotic potentials of culture medium, which thereby modifies cell wall possessions (extension, composition, and hardening process) (Pritchard et al. 1991). Sucrose is considered as an imperative source of carbohydrate and functions as an osmoticum in the culture medium (Hartmann et al. 1997), utilized in various energy requirement processes, and occurs at the pace of existing metabolic substrates intended for root introduction and development. Moreover, the sucrose stored in micropropagated plantlets helps in ex vitro acclimatization and physiological adaptations in TC-raised plants (Pospisilova et al. 1999). Comparable outcome in micropropagation of *W. somnifera* has been documented by Fatima et al. 2015.

9.3.5 Outcome of Different pH Levels

An optimal pH is requisite for growth and development of plant cells and tissue cultured in vitro.

It contributes significant position in uptake of nutrients, enzymatic, and hormonal actions (Bhatia and Ashwath 2005). pH controls cytoplasmic activities, which influence cell division and shoots growth in plants during morphogenesis, besides, it did not interfere in cells membrane functions and cytoplasm's buffered pH. Hence, the effect of different medium pH (4.2, 5.2, 5.8, 6.4, and 7.0) was also examined on MS medium having the optimal concentrations standardized. The highest shoots quantity and length (15.0 ± 0.00, 2.10 ± 0.10 cm) were harvested at pH 5.8, and thus was, considered as an optimum pH for maximum multiplication and elongation of shoots (Fig. 9.4).

The adjustment in pH of the medium could comprise different consequences that could manipulate performance and explants expansion (George et al. 2008). The modifications in pH of cells, tissues, and organs are owing to the ions absorbed from the medium. Furthermore, pH also manipulates the position of the solidifying agent in a medium. It was observed that higher pH level (above 6) fabricates hard medium, while a pH below 5 does not solidify the medium adequately (Bhatia and Ashwath 2005). Medium pH also influences the status of solidifying agent (Bhatia and Ashwath 2005). pH below optimal level resulted in watery medium, slight acidic in nature declines the number of shoots. Whereas, high (6.0) pH level resulted in shoots malformation and formation of callus at the base of cultures. Our findings substantiate with earlier findings reported by Fatima et al. 2015.

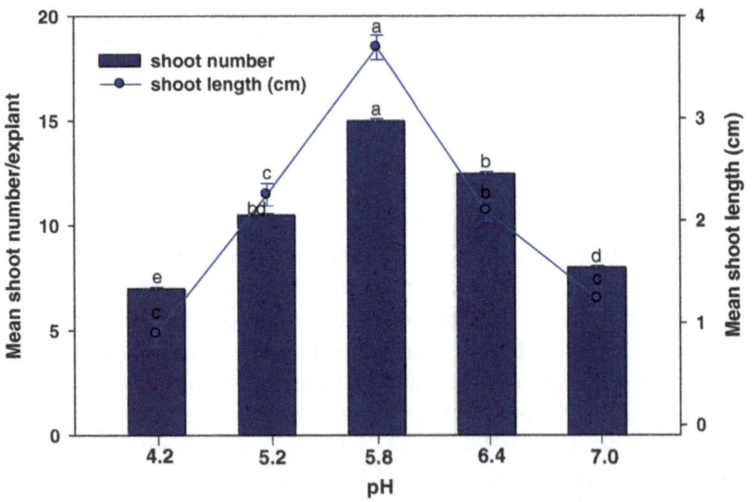

Fig. 9.4 Effect of medium pH on shoot proliferation from cotyledonary node explants supplemented with BA (2.5 μM) and NAA (0.5 μM) after 4 weeks of culture. Bars represent means ± SE. Bars denoted by the same letter within response variables are not significantly different ($P = 0.05$) using Duncan's multiple range test

Fig. 9.5 (**a**) Culture showing shoots bud emergence and multiple shoot formation from cotyledonary node explants of *W. somnifera* on MS + BA (2.5 μM), after 4 weeks of culture condition. (**b**) Multiplication and proliferation of shoots on MS + BA (2.5 μM) + NAA (0.5 μM), after 8 weeks of culture. (**c**) In vitro rooting in micro-shoots on MS medium fortified with NAA (0.5 μM), after 4 weeks. (**d**) Well established *W. somnifera* plantlets after 3 months of transfer to field conditions

9.3.6 In Vitro Rooting and Acclimatization

Roots play an essential arrangement for the growth and buildup of water and nutrients supply to plants from the atmosphere. For *W. somnifera*, the in vitro rooting of micropropagated shoots was accomplished by auxin treatments. The rhizhogenesis occurs through induction, initiation, and expression phases. Auxins are significantly and primarily essential for these developmental stages of rooting. The maximum figure of roots (12.6 ± 0.10) and length of roots (3.58 ± 0.16 cm) per shootlet has been stimulated on MS medium fortified with NAA (0.5 μM), after 4 weeks of culture condition (Table 9.5; Fig. 9.5c). Rooting in vitro has been effectively employed for the organization of root system in numerous medicinal plants species including *W. somnifera* (Fatima et al. 2015); *Syzgium cumini* (Naaz et al. 2018); *Allamanda cathatica* (Khanam and Anis 2018), and so on.

In vitro raised plantlets have heterotrophic mode of nutrition and therefore they lack proper adjustment to the external atmospheric conditions. The micropropagated plants are kept in hardening chamber during lab to land transfer. There is a gradual series of acclimatization weeks in shades followed by lowering of air humidity (Pospisilova et al. 1998). During the 4 weeks period, plantlets having expanded leaves, healthy grown roots were effectively hardened off within the growth room. The acclimatization procedure thereby facilitated the successful transfer of TC-raised plantlets to soil (Fig. 9.5d).

Table 9.5 Effect of different auxins on in vitro root induction from microshoots of *W. somnifera* after 4 weeks of culture

Plant growth regulators (μM)					
IAA	IBA	NAA	Regeneration (%)	Mean number of roots	Mean root length (cm)
0.0			00	0.00 ± 0.00^{j}	0.00 ± 0.00^{h}
0.1			85	3.38 ± 0.16^{i}	$1.68 \pm 0.35^{f,g}$
0.5			85	7.30 ± 0.12^{e}	$2.76 \pm 0.25^{b,c}$
1.0			75	5.38 ± 0.14^{g}	$2.30 \pm 0.16^{c,d,e}$
2.0			70	4.30 ± 0.12^{h}	$1.80 \pm 0.08^{e,f,g}$
	0.1		65	$7.90 \pm 0.24^{c,d}$	$2.44 \pm 0.18^{c,d}$
	0.5		80	10.1 ± 0.18^{b}	3.04 ± 0.02^{b}
	1.0		80	$7.52 \pm 0.15^{d,e}$	$2.14 \pm 0.09^{d,e,f}$
	2.0		70	$5.38 \pm 0.16^{c,g}$	1.58 ± 0.11^{g}
		0.1	90	8.10 ± 0.18^{c}	$2.64 \pm 0.21^{b,c,d}$
		0.5	95	12.6 ± 0.10^{a}	3.58 ± 0.16^{a}
		1.0	85	10.0 ± 0.22^{b}	$2.30 \pm 0.12^{c,d,e}$
		2.0	70	6.50 ± 0.27^{f}	$1.90 \pm 0.06^{e,f,g}$

Values represent means ± standard error of 20 replicates per treatment in three repeated experiments. Means followed by the same letter are not significantly different (P= 0.05) using Duncan's multiple range test

9.4 Conclusion

In recent years, ayurveda has gathered increasing recognition of Ashwagandha for preventive and therapeutic properties, as the chemically manufactured preparations (drugs) for single target treatment prove limited achievement and unfavorable toxic effects. The proposed protocol for en masse propagation of *W. somnifera* from CN explants is found to be more promising, enviable, and remains the most consistent approach for large-scale propagation of true-to-type progenies. Moreover, a direct regeneration protocol established in plant represents a stable platform for conservation strategies, sustainable utilization of exciting germplasm, broadening the genetic base, genetic transformation, secondary metabolites production, and so on, in a short duration and limited space. The decisive function of various factors like PGRs, media types, carbon sources, and so on, at molecular, cellular, or organ/whole plant levels can be documented during the regulation of morphogenesis and manipulated for the improved and enhanced plant propagation and establishment.

Acknowledgments The financial assistance in the form of UGC-Post Doctoral Fellowship for Women Candidates (UGC-PDFWM 2017-18) to Dr. Nigar Fatima funded by University Grants Commission (UGC), Government of India, New Delhi, is deeply acknowledged. Moreover, the award of UGC-BSR Faculty Fellowship (2017) by The University Grants Commission, New Delhi to Prof. Mohammad Anis is greatly acknowledged.

References

Ahmad N, Anis M (2010) An efficient in vitro process for recurrent production of cloned plants of *Vitex negundo* L. Eur J For Res 130:135–144

Anderson U, Levinsh G (2008) Medium pH affects regeneration capacity and oxidative enzyme activity of *Pinus sylvestris* in tissue culture. Acta Univ Latviens Biol 745:25–35

Anis M, Ahmad N (2016) Plant tissue culture: propagation, conservation and crop improvement. Springer, Singapore, pp 3–13

Bharti SK, Bharti A, Tewari SK, Sindhu OP, Roy R (2011) Application of HR-MASNMR spectroscopy for studying chemotype variation of *Withania somnifera* (L.) Dunal. Magn Reson Chem 49:659–667

Bhatia P, Ashwath N (2005) Effect of medium pH on shoot regeneration from cotyledonary explants of tomato. Biotechnology 4:7–10

Bhatia A, Bharti SK, Tewari SK, Sindhu OP, Roy R (2013) Metabolic profiling for studying chemotype variations in *Withania somnifera* L. Dunal fruits using GC-MS and NMR spectroscopy. Phytochemistry 93:105–115

Compton ME, Gray DJ, Hiebert E, Lin CM (1994) Microprojectile bombardment prior to co-cultivation with *Agrobacterium* improves GUS expression in watermelon cotyledons. In Vitro 30A:62

Dakah A, Zaid S, Suleiman M, Abbas S, Wink M (2014) In vitro propagation of the medicinal plant *Ziziphora tenuior* L. and evaluation of its antioxidant activity. Saudi J Biol Sci 21:317–323

De Klerk GJM, Calamar A (2002) Effect of sucrose on adventitious root regeneration in apple. Plant Cell Tissue Org Cult 70:207–212

Faisal M, Ahmad N, Anis M, Alatar AA, Qahtan A (2018) Auxin-cytokinin synergism in vitro for producing genetically stable plants of *Ruta graveolens* using shoot tip meristems. Saudi J Biol Sci 25:273–277

Farooqi AA, Sreeramu BS (2004) Cultivation of medicinal plants. University Press, India
Fatima N, Anis M (2012) Role of growth regulators on in vitro regeneration and histological analysis in Indian ginseng (*Withania somnifera* L.) Dunal. Physiol Mol Biol Plants 18:59–67
Fatima N, Ahmad N, Anis M (2012) In vitro propagation of *Cuphea procumbens* Orteg. and evaluation of genetic fidelity in plantlets using RAPD markers. J Plant Biochem Biotechnol 21:51–59
Fatima N, Ahmad N, Ahmad I, Anis M (2015) Interactive effects of growth regulators, carbon sources, pH on plant regeneration and assessment of genetic fidelity using single primer amplification reaction (SPARS) technique in *Withania somnifera* L. Appl Biochem Biotechnol 177:118–136
Gamborg OL, Miller RA, Ojima K (1968) Nutrient requirements of suspension cultures of soyabean root cells. Exp Cell Res 50:151–158
George EF, Hall MA, De Klerk GJ (2008) Plant propagation by tissue culture, The Background, vol 1, 3rd edn. Springer, Dordrecht
Hartmann HT, Kester DE, Davies FT, Geneva RL (1997) Plant propagation: principles and practices, 6th edn. Prentice-Hall of India Private Ltd, New Delhi, India
Khanam MN, Anis M (2018) Organogenesis and efficient in vitro plantlet regeneration from nodal segments of *Allamanda cathartica* L. using TDZ and ultrasound assisted extraction of quercetin. Plant Cell Tissue Organ Cult 134:241–250
Lipavaska H, Konradova H (2004) Somatic embryogenesis in conifers: the role of carbohydrate embolism. In Vitro Cell Develop Biol Plant 40:23–40
Lloyd GB, McCown BH (1981) Commercially feasible micropropagation of mountain laurel, *Kalmia latifolia* by use of shoot tip culture. Int Plant Propagat Soc Proc 30:421–427
Mahima A, Rahal R, Latheef SK, Samad HA (2012) Immunomodulatory and therapeutic potentials of herbal, traditional/indigenous and ethnoveterinary medicines. Pak J Biol 15:754–774
Moubayidin L, Di Mambro R, Sabatini S (2009) Cytokinin-auxin crosstalk. Trends Plant Sci 14:557–562
Müller D, Leyser O (2011) Auxin, cytokinin and the control of shoot branching. Ann Bot 107:1203–1212
Murashige T, Skoog F (1962) A revised medium for rapid growth and bioassays with tobacco tissue cultures. Physiol Plant 15:473–497
Naaz A, Shahzad A, Anis M (2018) Effect of adenine sulphate interaction of growth and development of shoot regeneration and inhibition of shoot tip necrosis under in vitro condition in *Syzygium cuminii* L.—a multipurpose tree. Appl Biochem Biotechnol 173:90–102
Nayak SA, Kumar S, Satapathy K, Moharana A, Behera B, Barik DP, Acharya L, Mohapatra PK, Jena PK, Naik SK (2012) In vitro plant regeneration from cotyledonay nodes of *Withania somnifera* (L.) dunal and assessment of clonal fidelity using RAPD and ISSR markers. Acta Physiol Plant 35:3–10
Patel P, Krishnamurthy R (2013) Feasibility studies on in vitro mass scale—propagation of Indian Ashwagandha (*Withania somnifera*) cultivars for commercial purpose. J Pharmacogn Phytochem 2:168–174
Phillips GC, Collins GB (1979) Genetic instability of plant tissue cultures: breakdown of normal controls. Crop Sci 19:59–64
Pospisilova J, Wilhelmova N, Synkova H, Catsky J, Krebs D, Tisha I, Hanackova B, Snopek J (1998) Acclimation of tobacco plantlets to ex vitro conditions as affected by application of abscisic acid. J Exp Bot 49:863–869
Pospisilova J, Ticha I, Kadlecek P, Haisel D, Plzakova S (1999) Acclimatization of micropropagated plants to ex vitro conditions. Biol Plant 42:481–497
Preece JE, Shutter EG (1991) Acclimatization of micropropagated plants to the greenhouse and fields. In: Debergh PC, Zimmerman RH (eds) Micropropagation: technology and application. Kluwer, Dordrecht, pp 71–93
Pritchard J, Wyn-Jones RG, Tomos AD (1991) Turgor, growth and rheological gradients in wheat roots following osmotic stress. J Exp Bot 42:1043–1049

Rout GR, Samantaray S, Das P (2000) In vitro manipulation and propagation of medicinal plants. Biotechnol Adv 18:91–120

Sen J, Sharma AK (1991) Micropropagation of *Withania somnifera* (L.) from germinating seeds and shoot tips. Plant Cell Tissue Org Cult 26:71–73

Shabir I, Wani H, Kumar V, Shriram V, Sah SK (2016) Phytohormones and their metabolic engineering for abiotic stress tolerance in crop plants. Crop J 4:162–176

Siddique I, Bukhari NAW, Siddique I (2015) Influence of plant growth regulators on in vitro shoot multiplication and plantlet formation in *Cassia angustifolia* Vahl. Braz Arch Biol Technol 58:686–691

Supe U, Dhote F, Roymon MG (2006) In vitro plant regeneration of *Withania somnifera*. Plant Tissue Cult Biotechnol 16:111–115

Tabei Y, Kanno T, Nishio T (1991) Regulation of organogenesis and somatic embryogenesis by auxin melon, *Cucurmis melo*. Plant Cell Rep 10:225–229

Tiwari R, Chakraborty S, Saminathan M, Dharma K, Singh SV (2014) Ashwagandha (*Withania somnifera*): role in safeguarding health, immnunomodulatory effects, combating infections and therapeutic applications: a review. J Biol Sci 14:77–94

Wickson M, Thimann KV (1958) The antagonism of auxin and kinetin in apical dominance. Plant Physiol 11:63–74

Williams RR, Taji AM, Winney KA (1990) The effect of *Ptilotus* plant tissue on pH of in vitro media. Plant Cell Tissue Org Cult 22:153–158

Yadav V, Malan L, Jaiswal VS (1990) Micropropagation of *Morus niger* from shoot tip and nodal explants of mature trees. Sci Hortic 31:61–64

Substitution of Benzyladenine with *meta*-Topolin Improved Shoot Regeneration and Rooting in *Wedelia chinensis* (Osbeck) Merr. and Quantitative Estimation of Flavonoids

10

Mehrun Nisha Khanam, Saad Bin Javed, Naseem Ahmad, and Neetu Singh

Abstract

A successful regenerative method was developed for *Wedelia chinensis* by using nodal segments as explant. Nodal segments were implanted on Murashige and Skoog (MS) (Physiol Plant, 15: 473–497, 1962) medium augmented with different combinations (0.5, 1.5, 2.5, 5.0, or 7.5 μM) of benzyladenine (BA) and *meta*-Topolin (*m*T) singly as well as in combination with various auxins *viz*. Indole-3-acetic acid (IAA), indole-3 butyric acid (IBA), and α-naphthalene acetic acid (NAA) at various concentrations (0.1, 0.5, or 1.0 μM). MS medium supplemented with *m*T (2.5 μM) was most efficient in inducing axillary shoot induction and growth, producing 12.40 ± 0.22 average number of shoot with 5.39 ± 0.09 cm shoot length after 8 weeks of incubation. The highest regeneration frequency (90%) was achieved by the inclusion of low concentration of NAA (0.5 μM) with optimal *m*T (2.5 μM) in MS medium, as it produced 28.90 ± 0.27 mean number of shoots with an average of 6.01 ± 0.04 cm shoot length after 12 weeks of incubation. Further microshoots initiated rooting on ½ MS liquid medium containing 0.5 μM IBA which produced 3.60 ± 0.16 mean number of roots/shoot with 4.20 ± 0.08 cm root length after 4 weeks of incubation. In vitro raised plants with well-developed roots were acclimatized well in thermocol cups comprising sterile soilrite for 4 weeks under culture conditions. Significant enhancement in the pigment content (Chlorophyll a, b, and carotenoid) was observed during successive period (0, 7, 14, 21, and 28 days) of acclimatization. The total flavonoid content was highest in the cultures treated with *m*T (2.5 μM) + NAA (0.5 μM) producing 111.60 CE/g of flavonoid.

Keywords

Auxin · Cytokinin · Flavonoids · Micropropagation

M. N. Khanam (✉) · S. B. Javed · N. Ahmad · N. Singh
Plant Biotechnology Laboratory, Department of Botany, Aligarh Muslim University, Aligarh, India

I. Siddique (ed.), *Propagation and Genetic Manipulation of Plants*,
https://doi.org/10.1007/978-981-15-7736-9_10

10.1 Introduction

Considering the need of mankind from time immemorial, plants were the ultimate source of food, flavors, and medicine, and so on. There are many strong evidences regarding the use of plants for their medicinal values by Egyptians, Romans, and Chinese in ancient times (Cowan 1999). The use of herbal medicines is growing in the developed countries, pharmaceutical industries directly or indirectly use about 40% of compounds (secondary metabolites) obtained from plants like tannins, alkaloids, flavonoids, saponins, steroids, and so on (Rout et al. 2000) because chemically synthesized compounds are expensive than plant-based compounds (Sidhu 2011). Various health advantageous characteristics of nutritional flavonoids are identified for humans, their antiproliferative and antioxidant properties could shield the body from numerous ailments, such as cancer, inflammation, and cardiovascular disease (Nijveldt et al. 2001). Consequently, due to overexploitation of large number of medicinal plants for the extraction of secondary metabolites, some are under threat of extinction (Vines 2004; Edwards 2004). One such medicinally important species is *Wedelia chinensis* (Asteraceae), generally known as "Pilabhangara or Bringaraj." It is widely distributed in India, China, and Japan. In India, it is mainly found in the regions of Uttar Pradesh, Assam, and Arunachal Pradesh. *W. chinensis* is a perennial herb with light camphor-like odor (Agarwala et al. 2010). The plant contains acylated active constituents like germacrene, eudesmanoids, α-humulene, caryophyllene, squalene, phellandrene, *p*-cymene, sitosterol, wedelia-seco-kaurenolide, and leaves are rich in flavonoids, isoflavonoids, and wedelolactone (Rastogi and Mehrotra 1993). It was reported that the active constituents of *W. chinensis* possess suppression activity against androgen, liver damage, and development in prostate cancer cells (Lin et al. 2007). The root extracts of *W. chinensis* are utilized for tattooing, tinting gray hair, stimulating growth of hair, and has antihepatotoxic property (Lin et al. 1994). The leaf extract is used to treat cough, arthritis, gouty arthritis, cephalagia, uterine hemorrhage, menorrhagia, and skin diseases like alopecia (Meena et al. 2011; Chopra et al. 1956; Kirtikar and Basu 1975). The growth rate of Ehrlich's ascites carcinoma is inhibited by 5% ethanolic extract of *W. chinensis* plant (CSIR 1992). *W. chinensis* is conventionally propagated through seeds and stem cuttings. Being a cross-pollinated species, it shows high level of variations among seed raised progeny and thus varies in quantity and quality of active constituents (Martin et al. 2003). The collections of plants from the wild have several disadvantages. Selective harvesting of a particular species leads to the ecological imbalance in that area; moreover, there are high chances of contamination as most of the time personnel involved in collection are not experts in species identification.

Therefore, plant tissue culture technique provides an unconventional mode of plant production in large number within much shorter duration to meet the local and international demand. The technique involves the interaction of various PGRs with different types of explants for direct and indirect organogenesis. The cytokinin–auxin interaction, ratio, and concentration determine the shoot and root regeneration from explants (Hoque 2010). Recently, some investigators have reported the dominance of an advanced class of aromatic cytokinin with hydroxylated benzyl ring (topolins) in plant tissue culture (Aremu et al. 2012). The potency of *meta*-topolin (*m*T) as compared to 6-benzyladenine (BA) on shoot multiplication was investigated. The synergistic outcome of auxins with optimum cytokinin concentration in shoot proliferation was also examined.

The available reports describing micropropagation of *W. chinensis* use BA as the cytokinin supplement (Martin et al. 2003; Agarwala et al. 2010). BA has been implicated to induce various physiological disorders causing problem in rooting and acclimatization of the regenerated shoots (Werbrouck et al. 1995). Alternatively, various analogs of BA like the Topolins can prove more effective substitute for BA. Therefore, *m*T can be used to develop much more efficient and improved protocol. However, only limited workers have explored its use and none in the case of *W. chinensis*.

The present study was designed to explore the morphogenic response of *m*T so as to assess the potency of *m*T as compared to BA on shoot multiplication for developing efficient micropropagation system for *W. chinensis* using nodal explants, physiological parameters, and estimate total flavonoids from the leaves of regenerated plantlets.

10.2 Materials and Methods

10.2.1 Explant Collection

Mature nodal explants of young shoots (first three nodes) of *W. chinensis* were harvested from a 2-year-old plant, maintained in the botanical garden of the University, Aligarh, India.

10.2.2 Explant Preparation

The young shoots comprising two to three nodes were washed for 15 min directly below the tap water followed by the application of 5% (v/v) labolene (Qualigens, India) for 5 min. Explants were again cleaned three to four times with distilled water, treated with 1% (v/v) Savlon for 1 min to remove adherent infectants, then washed again with sterilized distilled water. Explants were further externally disinfected in the laminar hood with 0.1% (w/v) mercuric chloride ($HgCl_2$) for 2.0 min followed by three to five times washing with sterilized distilled water. Thereafter, the explants (1.0–1.5 cm) were excised from the shoots and inoculated on shoot bud initiation medium.

10.2.3 Media and Culture Condition

Murashige and Skoog (MS 1962) nutrient media with 3% (w/v) sucrose (Qualigens, India) augmented with distinct concentrations of BA and *m*T (0.5, 1.5, 2.5, 5.0 or 7.5 μM) alone or in combination with various auxins *viz*. Indole-3-acetic acid (IAA), indole-3-butyric acid (IBA), or α-naphthalene acetic acid (NAA)) having different concentrations (0.1, 0.5, or 1.0 μM) were used. The pH of the medium was regulated to 5.8 using 1 N NaOH or 1 N HCl and then gelled with 0.8% (w/v) bacteriological grade agar prior to autoclaving at 121 °C and 15 psi for 18 min. MS medium lacking any plant growth regulators (PGRs) was used as control. All the cultures were maintained at 25 ± 2 °C under 16/8 h light and dark photoperiod in culture room, with a light intensity of 50 μmol m^{-2} s^{-1} and 55% relative humidity.

10.2.4 *In Vitro* Rooting and Acclimatization

In vitro generated healthy shoots (4–5 cm) were removed carefully and transferred onto root initiation media containing of full and half-strength MS liquid medium supplemented with various auxins (IAA, IBA, or NAA) at different concentrations (0.1, 0.5, and 1.0 μM). Data were documented on percentage response, root number, and length after 4 weeks of incubation. Well-grown rooted plants were cleaned directly with tap water to clean away the adherent media and transplanted to thermocol containers consisting of Soilrite™ (Keltech Pvt. Ltd., Bangalore). The plastic glass was then capped with plastic covers to maintain relative humidity in culture room condition and watered every 3 days. The plastic covers were removed after 2 weeks so that it accustom the plants to land requirements. After the 4 week, the plants were transplanted to earthen pots comprising garden soil and kept in a greenhouse.

10.2.5 Chlorophyll Estimation

Chlorophyll (a and b) and carotenoids in leaf tissue were estimated using the McKinney (1941) method. About 1 g of fresh leaf tissues from interveinal area was taken after 0 (juvenile leaves), 7, 14, 21, and 28 (fully expanded and mature leaves) days of hardening. The leaves were ground in 5 mL 80% acetone using motar and pestle. The extract was filtered using Whatman No. 1 filter paper. Total filtrate was collected in a test tube and final volume made 10 mL using 80% acetone. The optical density (OD) of chlorophyll was read at 645 and 663 nm wavelength, while for carotenoid, OD was recorded at 480 and 510 nm wavelength using UV–spectrophotometer (UV-Pharma Spec1700, Shimadzu, Japan).

10.2.6 Preparation of Plant Extract

The leaves from the control and regenerated plants and control (untreated) were air-dried separately, powdered to moderately coarse powder with the help of mortar and pestle. Two gram materials from each sample were extracted with 20 mL ethanol by soaking the samples overnight into it. Later, the samples were warmed in water bath at 45 °C, filtered using Whatman filter paper No. 1. The collected filtrate was concentrated under vacuum.

10.2.6.1 Estimation of Total Flavonoid (TF)

Aluminum chloride colorimetric assay was employed to measure total flavonoid content in the plant extract (Zhishen et al. 1999). Initially, measuring flask (10 mL) consisting 4 mL of distilled water was added with 1 mL of leaf extracts and standard solution of catechin (1 mg/mL), to this solution 0.3 mL 5% $NaNO_2$ was added. After 5 min, 0.3 mL solution of 10% $AlCl_3 \cdot 6H_2O$ was added into volumetric flask (10 mL). Thereafter 1 min, 2 mL of 1 M NaOH was added and the total volume was made upto 10 mL using distilled water. After thorough mixing the absorbance was measured against prepared reagent blank at 510 nm. Total flavonoid content of leaf extracts expressed as mg catechin equivalents (CE)/g of extract. All samples were analyzed in triplicates.

10.2.7 Statistical Analysis

The whole experimentation was set up in randomized block design and replicated thrice with ten replicates for each treatment. The data were subjected to analysis of variance (ANOVA) to differentiate between means using SPSS Ver 16 (SPSS Inc., Chicago, USA). All the results were articulated in mean ± standard error.

10.3 Results and Discussion

10.3.1 Effect of BA and *m*T on shoot induction

In plant tissue culture, the selection and use of cytokinin determine the efficiency in inducing shoot multiplication and growth. BA is used in many micropropagation systems because of its being economical and effective in shoot regeneration in many plant species. However use of BA also has some disadvantages like negative impact on rooting and acclimatization (Werbrouck et al. 1995). BA-derived analogs such as *meta*-Topolin used as an alternative plant growth regulator to refrain the adverse effects of BA in cultures. In *W. chinensis, m*T treatments proved significantly better to induce shoot production when compared to control and BA treatment. After 8 weeks of culture, 2.5 µM *m*T produced higher shoot number 12.40 ± 0.22 with 5.39 ± 0.09 cm mean shoot length having 80.00 ± 0.29 percentage responsive explants as compared to BA, which even at higher concentration produced lesser

Table 10.1 Effect of BA and *m*T for shoot regeneration on nodal explants after 8 weeks of incubation

Cytokinins (μM)				
BA	*m*T	Percent response	Mean shoot number	Mean shoot length (cm)
0.0	0.0	0.00 ± 0.00^{i}	0.00 ± 0.00^{i}	0.00 ± 0.00^{g}
0.5		45.90 ± 0.27^{h}	1.80 ± 0.13^{h}	1.55 ± 0.11^{f}
1.5		55.60 ± 0.22^{f}	3.60 ± 0.16^{f}	2.80 ± 0.08^{e}
2.5		66.90 ± 0.37^{d}	6.00 ± 0.21^{c}	4.05 ± 0.11^{c}
5.0		75.20 ± 0.24^{b}	8.40 ± 0.16^{b}	5.03 ± 0.04^{b}
7.5		59.90 ± 0.34^{e}	4.50 ± 0.16^{e}	3.50 ± 0.16^{d}
	0.5	59.30 ± 0.26^{e}	5.11 ± 0.17^{d}	2.65 ± 0.07^{e}
	1.5	74.50 ± 0.68^{b}	8.00 ± 0.25^{b}	3.08 ± 0.08^{c}
	2.5	80.00 ± 0.29^{a}	12.40 ± 0.22^{a}	5.39 ± 0.09^{a}
	5.0	70.30 ± 0.21^{c}	6.00 ± 0.29^{c}	4.10 ± 0.06^{c}
	7.5	49.20 ± 0.29^{g}	2.60 ± 0.16^{g}	2.50 ± 0.16^{e}

Values represent means ± standard error of readings of ten replicates per treatment in three repeated experiments. Means sharing the same letter are not significantly different ($P = 0.05$) using Duncan's multiple range test

number of shoots (Table 10.1; Fig. 10.1a). Similar findings of BA and *m*T effect on cultures were reported in *Spathiphyllum floribundum* (Werbrouck et al. 1996). The findings of their comparative study were that the key metabolite of *m*T was more stable and balanced than the main metabolite of BA, because BA had a negative influence on rooting and acclimatization as the *O*-glucoside does not degrade rapidly during acclimatization. The results showed that plants treated with *m*T generated considerably maximum root number and higher root length than those treated with BA. Limited reports are available on the use of topolins indicating that this group of cytokinins could be a new source of cytokinins with high morphogenetic activity. Kaminek et al. (1987) reported that *m*T was nearly twice as active and efficient as BA in the initiation of shoot growth in wheat plant species.

10.3.2 Effect of Cytokinin: Auxin Combination on Shoot Proliferation

The additive or interdependent influence of auxins with cytokinins on proliferation of shoots has been studied in numerous plant species (Shimizu-Sato et al. 2009). The equilibrium among auxin and cytokinin is an important regulator of in vitro organogenesis, demonstrated in the classical experiment of Skoog and Miller (1957). Among the three auxins used, NAA does not involve active uptake to pass through the plasma membrane into cells (Nordstrom et al. 2004). The optimum concentration of BA (5.0 μM) and *m*T (2.5 μM) was tested with numerous concentrations (0.1, 0.5, or 1.0 μM) of auxins. Among the tested concentration, *m*T (2.5 μM) + NAA (0.5 μM) significantly increased shoot production with 90% responsive explants, produced 28.90 ± 0.27 shoot number with 6.01 ± 0.04 cm shoot length after

Fig. 10.1 (**a**) Multiple shoot induction from nodal explant cultured on MS medium augmented with *m*T (2.5 μM) after 8 weeks of culture. (**b**) Shoot proliferation and multiplication on MS medium augmented with BA (5.0 μM) + NAA (0.5 μM) after 12 weeks of incubation. (**c**) Mass multiplication on MS + *m*T (2.5 μM) + NAA (0.5 μM) after 12 weeks of incubation. (**d**) Rooting in the regenerated microshoots on ½ MS liquid media containing 0.5 μM IBA after 4 weeks of incubation

Table 10.2 Effect of various auxins supplemented with optimized *m*T (2.5 μM) in MS medium on nodal explants after 12 weeks of incubation

Auxins (μM)					
IAA	IBA	NAA	Percent response	Mean shoot number	Mean shoot length (cm)
0.00	0.00	0.00	0.00 ± 0.00^{j}	0.00 ± 0.00^{i}	0.00 ± 0.00^{j}
0.1			72.40 ± 0.68^{e}	15.00 ± 0.21^{d}	3.85 ± 0.03^{e}
0.5			82.10 ± 0.60^{c}	21.60 ± 0.88^{b}	4.50 ± 0.03^{c}
1.0			40.60 ± 0.22^{i}	10.60 ± 0.16^{f}	2.82 ± 0.04^{h}
	0.1		74.10 ± 0.27^{d}	12.70 ± 0.15^{e}	3.63 ± 0.04^{f}
	0.5		65.50 ± 0.37^{g}	6.70 ± 0.15^{g}	2.97 ± 0.05^{g}
	1.0		54.60 ± 0.33^{h}	4.40 ± 0.16^{h}	2.04 ± 0.01^{i}
		0.1	84.60 ± 0.38^{b}	21.00 ± 0.25^{b}	5.00 ± 0.03^{b}
		0.5	90.00 ± 0.47^{a}	28.90 ± 0.27^{a}	6.01 ± 0.04^{a}
		1.0	70.00 ± 0.47^{f}	16.90 ± 0.27^{c}	4.15 ± 0.01^{d}

Values represent means ± standard error of readings of ten replicates per treatment in three repeated experiments. Means sharing the same letter are not significantly different ($P = 0.05$) using Duncan's multiple range test

Table 10.3 Effect of various auxins supplemented with optimized BA (5.0 μM) in MS medium on nodal explants after 12 weeks of incubation

Auxins (μM)					
IAA	IBA	NAA	Percent response	Mean shoot number	Mean shoot length (cm)
0.00	0.00	0.00	0.00 ± 0.00^{i}	0.00 ± 0.00^{i}	0.00 ± 0.00^{i}
0.1			60.00 ± 0.63^{e}	6.70 ± 0.21^{f}	3.16 ± 0.08^{d}
0.5			70.00 ± 0.47^{c}	11.50 ± 0.34^{c}	$4.10 \pm 0.06^{b,c}$
1.0			53.00 ± 0.57^{g}	3.30 ± 0.21^{h}	2.67 ± 0.03^{e}
	0.1		60.00 ± 0.51^{e}	7.40 ± 0.22^{e}	4.30 ± 0.08^{b}
	0.5		55.00 ± 0.36^{f}	4.60 ± 0.16^{g}	3.25 ± 0.08^{d}
	1.0		50.00 ± 0.63^{h}	3.00 ± 0.21^{h}	2.20 ± 0.08^{f}
		0.1	76.00 ± 0.47^{b}	14.50 ± 0.16^{b}	3.97 ± 0.05^{c}
		0.5	84.00 ± 0.24^{a}	17.30 ± 0.26^{a}	5.23 ± 0.05^{a}
		1.0	65.00 ± 0.47^{d}	10.40 ± 0.22^{d}	2.50 ± 0.08^{e}

Values represent means ± standard error of readings of ten replicates per treatment in three repeated experiments. Means sharing the same letter are not significantly different ($P = 0.05$) using Duncan's multiple range test

12 weeks of culture (Table 10.2; Fig. 10.1c), signifying a synergistic effect of NAA on shoot proliferation. Other *m*T–auxins combinations proved less responsive. However, among BA (5.0 μM) + NAA (0.5 μM) combinations could only produce a maximum of 17.30 ± 0.26 mean shoots and 5.23 ± 0.05 cm shoot length in 84% cultures after 12 weeks of incubation (Table 10.3; Fig. 10.1b).

10.3.3 Rooting in Microshoots

Regenerated microshoots (5.0 cm) from nodal explants did not induce any root on full or ½ strength MS media devoid of plant growth regulators. However, the cut ends of microshoots instigated root primordial when cultured on ½ MS liquid medium augmented with auxins after 2 weeks of incubation. IBA (0.5 μM) was most persuasive in inducing rhizogenesis in 94% cultures which produced about 3.60 roots, with an average root length of 4.20 ± 0.08 cm per shoot (Table 10.4; Fig. 10.1d). The effectiveness of treating microshoots with IBA for root induction has been reported in several plant species, *Saussurea obvallata* (Dhar and Joshi 2005), *Aspilia montevidensi* (Kunst-Barosky et al. 2011), and *Grindelia chiloensis* (Wassner and Ravetta 2000). However, NAA and IAA with ½ MS media produced basal callusing at the cut end and produced lesser number of roots than IBA. Hence, IBA was found to be superior for root initiation without any induction of callus in *W. chinensis*.

10.3.4 Acclimatization

Acclimatization is the most essential step in tissue culture. Therefore, well-developed healthy rooted plantlets with six to seven leaves were transferred to pots containing sterile Soilrite™ for 4 weeks. The plantlets were positively transferred to greenhouse after acclimatization where of 94% survival rate was recorded (Fig. 10.2). All the regenerated plantlets showed normal morphology and growth similar to that of the mother plant.

Table 10.4 Effect of ½ MS medium augmented with various auxins on rooting in microshoots after 4 weeks of incubation

Auxins (μM)					
IAA	IBA	NAA	Percent response	Mean root number	Mean root length (cm)
00			0.00 ± 0.00^{j}	0.00 ± 0.00^{f}	0.00 ± 0.00^{f}
0.1			61.10 ± 0.31^{f}	0.60 ± 0.16^{e}	0.90 ± 0.25^{e}
0.5			67.70 ± 0.15^{d}	1.60 ± 0.16^{b,c}	1.85 ± 0.41^{c,d}
1.0			51.00 ± 0.25^{h}	0.50 ± 0.16^{e}	0.70 ± 0.26^{e}
	0.1		85.00 ± 0.25^{b}	2.10 ± 0.23^{b}	3.25 ± 0.08^{b}
	0.5		94.90 ± 0.23^{a}	3.60 ± 0.16^{a}	4.20 ± 0.08^{a}
	1.0		75.30 ± 0.21^{c}	1.50 ± 0.16^{b,c}	2.40 ± 0.12^{c}
		0.1	56.00 ± 0.24^{g}	0.90 ± 0.31^{c,d}	0.90 ± 0.31^{e}
		0.5	64.20 ± 0.35^{e}	1.50 ± 0.37^{b,c}	1.55 ± 0.43^{d,e}
		1.0	41.40 ± 0.22^{i}	0.50 ± 0.22^{e}	0.80 ± 0.32^{e}

Values represent means ± standard error of readings of ten replicates per treatment in three repeated experiments. Means sharing the same letter are not significantly different ($P = 0.05$) using Duncan's multiple range test

Fig. 10.2 Acclimatized plantlet

10.3.5 Photosynthetic Pigments

The variation in pigment concentration (Chlorophyll and carotenoid contents) was assessed in regenerated plants when transferred from in vitro to ex vitro. During the initial period of acclimatization a reduction in carotenoid and chlorophyll a and b was recorded. The pigment contents increased significantly with an increase in acclimatization period. According to Borkowska (2001), the young leaves of the in vitro raised strawberry plants acclimatized to ex vitro condition exhibited low phytochemical activity but as the period of acclimatization progresses with time, these pigments became more active in mature leaves. In the present study, the Chl a and b content were low (0.25 $\pm$ 0.04) mg g^{-1} and (0.21 $\pm$ 0.01) mg g^{-1} on 0 days of acclimatization, got increased up to (0.42 $\pm$ 0.02) mg g^{-1} and (0.34 $\pm$ 0.02) mg g^{-1} after 21 days and reached maximum (0.45 $\pm$ 0.05) mg g^{-1} and (0.37 $\pm$ 0.06) mg g^{-1} after 28 days of transfer (Fig. 10.3). Initial reduction in chlorophyll content may be due to poorly developed chloroplast and disorganized grana. Enhancement in

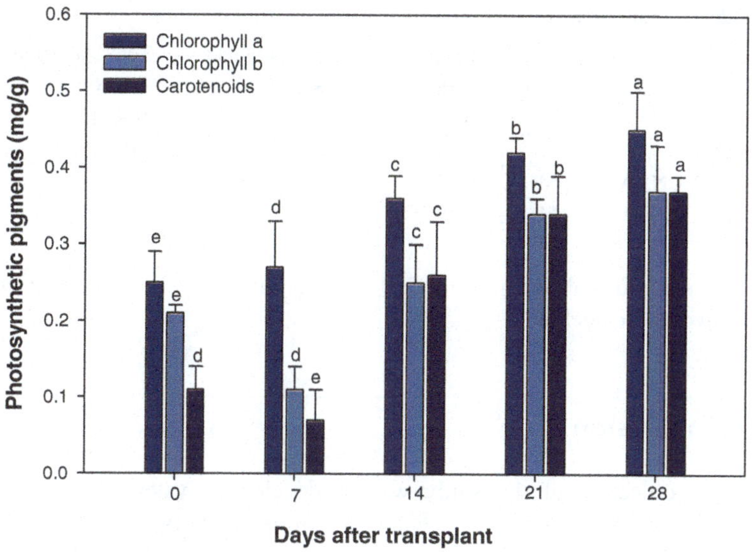

Fig. 10.3 Net photosynthetic content

pigment contents is attributed to the induction of chlorophyll synthesis enzyme required for chlorophyll biosynthesis. Similar results have been reported by Posposilova et al. (1999) and Jeon et al. (2005).

Carotenoid plays an important role in photoprotection of chlorophyll pigments under stress (Kenneth et al. 2000) which might be engendered during the acclimatization period. The carotenoid contents enhanced progressively during the transplantation period. The highest concentration (0.37 ± 0.02) mg g^{-1} was recorded after 28 days of acclimatization (Fig. 10.3). Increase in carotenoid levels is not unexpected as carotenoids are reported to be involved in protecting the photosynthetic machinery from photo-oxidative damage (Jeon et al. 2005).

10.3.6 Estimation of Flavonoids

Two grams of leaf sample collected from regenerated plantlets were screened for estimation of flavonoids. The plantlets grown on MS + *m*T (2.5 μM) + NAA (2.5 μM) gave highest yield (0.19 g or 9.5%) (Table 10.5). The method proposed by Mishra et al. (2011) and Yang et al. (2008) for extraction of flavonoids was used.

Aluminum chloride colorimetric method using spectrophotometer was employed to estimate total flavonoid content in leaf extract (Table 10.5) and expressed in milligram catechin equivalent to per gram of the extract. The content varied among treated samples and highest content (111.60 mg CE/g) was recorded in leaf tissue of plantlets raised on MS + *m*T (2.5 μM) + NAA (2.5 μM). Colorimetric reactions are speedy, easy to perform, and widely used to determine total flavonoids content in medicinal plant. This is based on the formation of the flavonoids–aluminum complex

Table 10.5 Quantification of total flavonoid content

Treatment	Extract (g)	Percent yield	Total flavonoids (CE)/g
Natural source	0.16	8.0	109.10
BA (5.0 μM)	0.17	8.5	109.51
*m*T (2.5 μM)	0.17	8.5	110.21
BA (5.0 μM) + NAA (0.5 μM)	0.18	9.0	110.73
*m*T (2.5 μM) + NAA (0.5 μM)	0.19	9.5	111.60

which has maximum absorptivity at 510 nm. The treated extract showed the presence of flavonoid catechin.

10.4 Conclusion

The overall objective of the study was to develop a system for an improved regeneration protocol of *W. chinensis* from axillary shoot buds. The micropropagation established multiplication year-round can be used for production of quality planting material at a faster rate than the available protocols in *W. chinensis*. The experiments have clearly indicated the potentiality of *m*T over BA for shoot multiplication and their health without causing any physiological disorders in rooting and acclimatization of the regenerated shoots. The in vitro culture strategies developed could be a useful tool to increase the biomass and yield of flavonoids accumulated in *W. chinensis* cultures.

References

Agarwala B, Azam FS, Khatun MA, Rahman F, Rahmatullah M (2010) Simultaneous shoot regeneration and rhizogenesis of *Wedelia chinensis* for in vitro clonal propagation. Am Eurasian J Sustain Agricult 4:65–69

Aremu AO, Bairu MW, Dolezal K, Finnie JF, Van Staden J (2012) Topolins: a panacea to plant tissue culture challenges. Plant Cell Tissue Org Cult 108:1–16

Borkowska B (2001) Morphological and physiological characteristics of micropropagated strawberry plants rooted in vitro or ex vitro. Sci Hortic 89:195–206

Chopra RN, Nayar SL, Chopra IC (1956) Glossary of Indian medicinal plants. CSIR, New Delhi

Cowan MM (1999) Plant products as antimicrobial agents. Clin Microbiol Rev 12:564–582

CSIR (1992) The useful plants of India (reprints). PID, New Delhi, India

Dhar U, Joshi M (2005) Efficient plant regeneration protocol through callus for *Saussurea obvallata* (DC.) Edgew. (Asteraceae): effect of explant type, age and plant growth regulators. Plant Cell Rep 24:195–200

Edwards R (2004) No remedy in sight for herbal ransack. New Scientist 181:10–11

Hoque ME (2010) In vitro tuberization in potato (*Solanum tuberosum* L.). Plant Omic J 3:7–11

Jeon MW, Ali MB, Hahn EJ, Paek KY (2005) Effects of photon flux density on the morphology, photosynthesis and growth of a CAM orchid, *Doritaenopsis* during post-micropropagation acclimatization. Plant Growth Regul 45:139–147

Kaminek M, Vanek T, Motyka V (1987) Cytokinin activities of N6-benzyladenosine derivatives hydroxylated on the side-chain phenyl ring. J Plant Growth Regul 6:113–120

Kenneth E, Pallet KE, Young AJ (2000) Carotenoids. In: Ruth GA, Hess JL (eds) Antioxidants in Higher Plants. CRC Press, Boca Raton, USA, pp 60–81
Kirtikar KR, Basu BD (1975) Indian medicinal plants, vol 2. M/s Bishen Singh Mahendrapal, New Delhi, India, p 1365
Kunst-Barosky T, Petry C, dos Santos SRF (2011) Rooting of *Aspilia montevidensis* (Spreng.) Kuntze (Asteraceae) Cuttings. VII Int Symp New Floricult Crops 1000:301–306
Lin SC, Lin CC, Lin YH, Shyuu SJ (1994) Hepatoprotective effects of Taiwan folk medicine: *Wedelia chinensis* on three hepatotoxin-induced hepatotoxicity. Medicine 22:155–168
Lin FM, Chen LR, Lin EH, Ke FC, Chen HY, Tsai MJ, Hsiao PW (2007) Compounds from *Wedelia chinensis* synergistically suppress androgen activity and growth in prostate cancer cells. Carcinogenesis 28:2521–2529
Martin KP, Benna MR, Joseph D (2003) High frequency axillary bud multiplication and ex vitrorooting of *Wedelia chinensis* (Osbeck) Merr.–A medicinal plant. Indian J Exp Biol 41:262–266
McKinney, G., 1941. Absorption of light by chlorophyll solutions. J. Biol. Chem. 140, 315–322
Meena AK, Rao MM, Meena RP, Panda P, Renu (2011) Pharmacological and phytochemical evidences for the plants of Wedelia Genus–a review. Asian J Pharm Res 1:7–12
Mishra G, Singh P, Garg VK, Parvez N, Yadav S, Hwisa N, Khosa RL (2011) Phytochemical screening and anticonvulsant activity of *Wedelia chinensis*. Int J Pharm Sci Res 2:25
Murashige T, Skoog F (1962) A revised medium for rapid growth and bioassays with tobacco tissue culture. Physiol Plant 15:473–497
Nijveldt RJ, van Nood E, van Hoorn DE, Boelens PG, van Norren K, van Leeuwen PA (2001) Flavonoids: a review of probable mechanisms of action and potential applications. Am J Clin Nutr 74:418–425
Nordstrom A, Tarkowski P, Tarkowska D, Norbaek R, Astotm C, Dolezal K, Sandberg G (2004) Auxin regulation of cytokinin biosynthesis in *Arabidopsis thaliana*: a factor of potential importance for auxin-cytokinin-regulated development. Proc Natl Acad Sci U S A 101:8039–8044
Posposilova J, Ticha I, Kadlecek P, Haisel D, Plzakova S (1999) Acclimatization of micropropagated plants to ex vitro conditions. Biol Plant 42:481–497
Rastogi RP, Mehrotra BN (1993) Compendium of Indian medicinal plants, vol 3. CDRI, Lucknow and Publication and Information Directorate, New Delhi, India, p 679
Rout GR, Samantaray S, Das P (2000) In vitro manipulation and propagation of medicinal plants. Biotechnol Adv 18:91–120
Shimizu-Sato S, Tanaka M, Mori H (2009) Auxin–cytokinin interactions in the control of shoot branching. Plant Mol Biol 69:429
Sidhu Y (2011) In vitro micropropagation of medicinal plants by tissue culture. Plymouth Student Scientist 4:432–449
Skoog F, Miller CO (1957) Symp Soc Exp Biol 11:118–131
Vines G (2004) Herbal Harvest with future: towards sustainable source for medicinal plants. Plantlife International, www.plantlife.co.uk
Wassner D, Ravetta D (2000) Vegetative propagation of *Grindelia chiloensis* (Asteraceae). Ind Crop Prod 11:7–10
Werbrouck SP, van der Jeugt B, Dewitte W, Prinsen E, Van Onckelen HA, Debergh PC (1995) The metabolism of benzyladenine in *Spathiphyllum floribundum* 'Schott Petite'in relation to acclimatisation problems. Plant Cell Rep 14:662–665
Werbrouck SP, Strnad M, Van Onckelen HA, Debergh PC (1996) Meta-topolin, an alternative to benzyladenine in tissue culture? Physiol Plant 98:291–297
Yang RY, Lin S, Kuo G (2008) Content and distribution of flavonoids among 91 edible plant species. Asia Pac J Clin Nutr 17:275–279
Zhishen J, Mengcheng T, Jianming W (1999) The determination of flavonoid contents in mulberry and their scavenging effects on superoxide radicals. Food Chem 64:555–559

Effect of Silicon Ions on In Vitro Morphogenesis in *Rauvolfia tetraphylla* Using Shoot Tip Explants

11

Sheikh Altaf Hussain

Abstract

The study evaluates the effect of Silicon ions on in vitro morphogenesis in *Rauvolfia tetraphylla* using aseptic shoot tip explants. The explants were cultured on MS medium supplemented with varied concentrations of BA either singly or in combination with NAA. Of the tested concentrations, best response (80%) in explants for in vitro shoot production was recorded on MS medium fortified by BA (5.0 μM) and NAA (0.5 μM) with mean shoot number (9.66 ± 0.57) and shoot length (3.76 ± 0.16 cm) after 8 weeks of incubation. However, on evaluation of the morphogenic effect of different concentrations of silicon ions with optimized levels of BA and NAA, an increase in the percentage response and regeneration frequencies were observed. Of the tested combinations, MS culture medium supplemented with (5.0 μM BA + 0.5 μM NAA + 5.0 mM Na_2SiO_3) yielded 100% morphogenic response with mean shoot number (15.33 ± 0.66) and average length (5.73 ± 0.26) after 8 weeks of culture. In vitro rhizogenesis in 93% microshoots with maximum mean number (9.66 ± 0.66) of roots with mean root length (3.56 ± 0.23) per shootlet was recorded on ½ WPM medium supplemented with 0.5 μM IBA, after 4 weeks of incubation. The plantlets regenerated via this protocol were successfully acclimatized and hardened off under controlled conditions of growth chamber before their plantation to field conditions. The plantlets grew well under open sun and 90% survival rate was recorded after 12 weeks of transfer. The protocol developed can be used for propagation and conservation of rare and endangered plant assets.

Keywords

Acclimatization · Benzyl adenine · In vitro rooting · Na_2SiO_3 · WPM medium

S. A. Hussain (✉)
Plant Biotechnology Laboratory, Department of Botany, Aligarh Muslim University, Aligarh, India

I. Siddique (ed.), *Propagation and Genetic Manipulation of Plants*,
https://doi.org/10.1007/978-981-15-7736-9_11

Abbreviations

BA	Benzyl adenine
cm	Centimetre
IBA	Indole-3-butyric acid
mM	millimole
Na_2SiO_3	Sodium meta-silicate
NAA	α-naphthalene acetic acid
μM	micro molar

11.1 Introduction

Rauvolfia tetraphylla is a potential medicinal shrub of family Apocynaceae and is commonly referred as devil pepper. The *R. tetraphylla* is a native species of tropical America and distributed in parts of Asia, Africa and India. The species is restricted to moist and warm parts of India such as Andhra Pradesh, Bihar, Howrah, Kerala, Madhya Pradesh, Orissa, Uttar Pradesh and West Bengal (Jyothi et al. 2012). Morphologically the plant body of *R. tetraphylla* is woody, evergreen and perennial and grows up to the height of 5 ft. Four unequal dark green leaves are found on each node, thus, the name *R. tetraphylla* (Anonymous 2003). It is potent source of many important alkaloids like ajmaline, alistonine, canescine, corynathene, desperdine, reserpine, tetraphyllicine and yohimbine and so on. The bioactive compound reserpine is used to cure nervous system disorders. Further it maintains blood glucose levels in diabetic patients and regulates normal blood pressure in humans (Hussain et al. 2018a). Root decoction is used for curing various intestinal infections like diarrhoea, dysentery and other helminthic diseases from pre historic times (Faisal et al. 2006; Hussain et al. 2018b). The species has been indiscriminately exploited from its natural habitat and sometime whole plants are uprooted to meet economic and medicinal gains for growing population and to satisfy increasing interest from pharmacological industries. Poor vegetative propagation methods, low seed germination and viability rates without any replantation programme have made this plant critically endangered (Faisal and Anis 2009). Therefore the need of hour is to find out the alternate propagation methods for it mass-scale replantation drive so that we can challenge all the growing demands. The in vitro micropropagation technique is the only way available for large-scale production, propagation and conservation of rare, elite and endangered plant species (Anis and Ahmad 2016). Hence, an attempt was successfully made to develop an efficient, reliable and economical in vitro regeneration protocol using aseptic nodal explants by employing tissue culture technology. The current study was undertaken to study role of silicon ions (most abundant element) on in vitro axillary shoot induction and proliferation. Silicon is known for its significant physiological role in enhancing the power of resistance under oxidative stresses caused by drought, salinity, metal toxicity and other plant diseases by modulating antioxidant enzymes (Sivanesan and Jeong 2014; Sivanesan et al. 2011).

11.2 Materials and Methods

11.2.1 Explants Source

The shoot tip explants (1.0 cm) were excised from 14 months old *R. tetraphylla* aseptic cultures which were maintained in plant biotech culture room of Aligarh Muslim University, Aligarh, India (Hussain et al. 2018a, b).

11.2.2 Inoculation and Culture Media

For in vitro shoot tip culture, the explants were cultured on Murashige and Skoogs (MS) medium (1962) enriched with various concentrations (1.0, 2.5, 5.0, 7.5 or 10.0 μM) of BA alone or in combination with various concentrations (0.1, 0.5, 1.0, 1.5 or 2.0 μM) of auxin NAA. The effect of sodium meta-silicate (Na_2SiO_3) at various concentrations (1.0, 2.5, 5.0, 7.5 or 10.0 mM) was also evaluated in the optimized culture media.

11.2.3 Culture Conditions

For all experiments, MS medium at pH 5.8, with enrichment of 3% sucrose as carbon source was used. The culture medium was autoclaved at 1.06 kg cm^2 pressure at 121 °C for 16 min after addition of 0.9% agar (Qualigens Pvt. Ltd., Mumbai, India) as solidifying agent. All inoculation and subculturing were done under sterile conditions inside laminar air flow hood and the cultures were maintained at 24 ± 2 °C for incubation in growth chamber under photoperiod of 16/8 h. The 40 W white fluorescent lamps provided desirable light intensity (60 μmol m^{-2} s^{-1}) and were periodically monitored by Lux photo metre. The relative humidity of the growth chamber was maintained at 60 ± 05% and was continuously monitored by hygrometer.

11.2.4 Subculturing

For further shoot proliferation, the cultures were subcultured after every 3 weeks onto a similar fresh culture medium.

11.2.5 In Vitro Rooting in Microshoots

Healthy microshoots (~5 cm) were excised and employed for in vitro root induction in full or half strength liquid Woody Plant Medium (WPM) supplemented with different concentrations (0.1, 0.25, 0.50, 0.75 or 1.0 μM) of IBA. The filter paper bridges made of Whatman filter paper number 1 were used for incubating the microshoots in rooting medium for 4 weeks.

11.2.6 Acclimatization and Hardening

At the end of 4 weeks, the regenerated mini plantlets were planted in polystyrene cups filled with autoclaved Soilrite™ (Keltech Energies Ltd., Bangalore) and were immediately covered by poly bags (transparent) in order to retain and ensure proper humidity around the plant body. On alternate days, the planting substrate Soilrite was sprayed with ½ MS stock solution without organic nutrients. In order for better acclimatization, the polymembrane cover was removed from the plantlets after 2 week of transfer. Thereafter, the regenerated plantlets were potted in the earthen pots containing 3:1 mixture garden soil and manure. After proper hardening, the pots were transferred and allowed to grow under field conditions.

11.2.7 Data Collection and Statistical Analysis

The whole micropropagation cycle was repeated thrice in order to attain the reproducibility of the regenerated protocol. Data were collected on every 4 week and were analysed by statistical software (SPSS *Ver*-16 Inc., Chicago, USA). The Duncan's Multiple Range Test (DMRT) was used for carrying out the significance difference among means.

11.3 Results and Discussion

11.3.1 Effect of Different Concentrations of BA on In Vitro Shoot Bud Induction

No morphogenic response was achieved in ST explants (shoot tip) on control medium (MS 1962 without PGR) even after 8 weeks. On augmentation of BA to the medium, the explants responded well for shoot initiation and multiplication. A differential response for axillary shoot bud initiation and differentiation was observed at varied concentrations. Of the tested concentrations, highest response (75%) with average mean shoot number (5.66 ± 0.88) and shoot length (3.46 ± 0.36 cm) per explant was recorded on MS medium fortified with 5.0 μM BA after 8 weeks of culture (Table 11.1). Similar observations have been reported in many woody species where BA was effective in promoting maximum shoot bud initiation and proliferation like *Euphorbia cotinifolia* (Perveen et al. 2013) and *Cryptolepis sanguinolenta* (Monney et al. 2016).

Table 11.1 Effect of BA, NAA and Na_2SiO_3 on in vitro shoot bud induction and proliferation on aseptic shoot tip explants of *R tetraphylla*, after 8 weeks of incubation

BA (μM)	NAA (μM)	Na_2SiO_3 (Mm)	Percentage response	Mean no of shoots/ explant	Mean shoot length (cm)
0.0	0.0	0.0	0.0	0.00 ± 0.00^{l}	$0.0\ 0 \pm 0.0^{j}$
1.0			33	1.66 ± 0.33^{k}	1.43 ± 0.23^{i}
2.5			49	$3.00 \pm 0.57^{i,j}$	2.24 ± 0.17^{h}
5.0			75	$5.66 \pm 0.88^{g,h}$	$3.46 \pm 0.36^{e,f}$
7.5			70	$4.33 \pm 0.88^{h,i}$	$2.93 \pm 0.22^{f,g}$
10.0			66	3.66 ± 0.33^{i}	2.20 ± 0.13^{h}
5.0	0.1		73	$6.33 \pm 0.33^{f,g}$	2.83 ± 0.08^{g}
–	0.5		80	9.66 ± 0.57^{d}	$3.76 \pm 0.16^{d,e}$
–	1.0		78	8.33 ± 0.66^{e}	3.58 ± 0.24^{f}
–	1.5		75	$7.66 \pm 0.33^{e,f}$	$3.36 \pm 0.14^{e,f}$
–	2.0		70	$6.66 \pm 0.33^{f,g}$	$3.66 \pm 0.13^{e,f}$
5.0	0.5	1.0	85	11.00 ± 0.57^{d}	$4.33 \pm 0.10^{c,d}$
–	–	2.5	88	13.66 ± 0.33^{b}	5.16 ± 0.38^{b}
–	–	5.0	100	15.33 ± 0.66^{a}	5.73 ± 0.26^{a}
–	–	7.5	95	$12.66 \pm 0.66^{b,c}$	$4.66 \pm 0.17^{b,c}$
–	–	10.0	90	$11.33 \pm 0.37^{c,d}$	$4.43 \pm 0.21^{c,d}$

Effect of MS medium augmented with different concentrations of BA, NAA and Na2Sio3 on in vitro shoot bud induction and proliferation on aseptic shoot tip explants of R. tetraphylla after 8 weeks of culture

11.3.2 Effect of Optimized Concentration of Benzyl-Adenine with Auxin NAA

On evaluation of the combined effect of the optimized concentration of BA with different concentrations of NAA, an enhanced shoot multiplication rate was observed. On accessing the combination effects, 80% explants responded on MS medium fortified with BA (5.0 μM) and NAA (0.5 μM) with maximum mean number (9.66 ± 0.57) of shoots having an average length of 3.76 ± 0.16 cm, after 8 weeks of incubation (Table 11.1). The synergistic outcome of higher level of BA with lower doses of NAA positively influenced the shoot regeneration frequencies by increasing the shoot number up to an optimal combination and thereafter a decreasing trend in shoot growth was recorded. The observed results are in accordance with prior reports (Ahmad et al. 2013; Perveen et al. 2013; Hussain et al. 2020) where BA and NAA combinations promoted higher shoot frequencies per explant than singly supplemented BA.

11.3.3 Effect of Silicon Ions with Optimized Concentrations of BA and NAA

On evaluation of morphogenic effect of various concentrations (1.0–10.0 Mm) of silicon ions in the form of sodium meta-silicate with optimized levels of BA and NAA, a synergistic increase shoot number was observed. Among tested combinations, MS medium augmented with (5.0 μM BA, 0.50 μM NAA and 5.0 mM Na_2SiO_3) yielded 100% response (Fig. 11.1a, b) and produced good quality shoots with a maximum of 15.33 ± 0.66 shoots per explant with an average length of 5.73 ± 0.26 cm, after 8 weeks of incubation (Table 11.1). The growth parameters increased up to the thresh hold level and above that retarded shoot growth was observed. All the cultures were healthier and possess larger internodes and leaves than BA or BA + NAA combination grown cultures. These results are in accordance with the results obtained in different plant species where Si has played a significant role for shoot growth and multiplication such as rice (Islam et al. 2005); reed (Mathe et al. 2012) and *Ajuga multiflora* (Sivanesan and Jeong 2014). Further no physiological disorder like hyperhydricity, shoot tip necrosis or shoot fasciation was observed in silicon mediated cultures even after 1 year while as in the BA or BA + NAA regenerated cultures these effects were evident only after 4 months of culture. The possible reason for the better growth in silicon-mediated cultures may be due to the maintenance of stability among different constituents of the culture medium by detoxifying the free radicals generated during the tissue culture process (Ma 2004; Prabagar et al. 2011; Sivanesan and Jeong 2014). Overall, our results indicate that in vitro morphogenic response and efficiency of micropropagation system depends upon the medium composition and their manipulation in right way.

11.3.4 Rooting of Regenerated Shoots

No root induction was observed on control medium, however, on enrichment of different concentrations of IBA promoted initiation of in vitro rooting after 1 week of incubation. The 93% microshoots rooted in ½ WPM enriched with 0.5 μM of IBA, and yielded maximum number (9.66 ± 0.66) of roots per shootlet with an average length of 3.56 ± 0.23 cm, after 4 weeks of incubation (Table 11.2). Half strength WPM medium in combination with IBA was found more effective for rhizogenesis (Fig. 11.1c). The results obtained are in corroboration with the established micropropagation protocols where IBA was found effective due to its stability and easy uptake by plant tissues for better in vitro rhizogenesis in plant species like *R. tetraphylla* (Hussain et al. 2018a, b) and *Allamanda cathartica* (Khanam and Anis 2018); *Tecoma stans* (Hussain et al. 2019a).

Fig. 11.1 (**a**) Shoot bud induction and multiplication on MS + (5.0 μM) BA + (0.5 μM) NAA + (5.0 mM) Na_2SiO_3 after 3 weeks of incubation. (**b**) Shoot proliferation after 8 weeks of incubation on MS + (5.0 μM) BA + (0.5 μM) NAA + (5.0 mM) Na_2SiO_3. (**c**) In vitro rooting in microshoots after 3 weeks of incubation on ½ WPM + 0.50 μM IBA (liquid medium). (**d**) Micropropagated plantlets after 4 weeks of acclimatization

Table 11.2 Effect of IBA on in vitro rooting in regenerated microshoots with full and half strength WPM after 4 weeks of incubation

IBA(μM)				
WPM	½ WPM	Percentage response	Mean root no/shootlet	Mean root length (cm)
0.00	0.0	00	0.00 ± 0.00^{g}	0.00 ± 0.00^{h}
0.10		53	1.33 ± 0.33^{f}	0.76 ± 0.13^{g}
0.25		67	3.66 ± 0.33^{e}	$1.26 \pm 0.12^{e,f}$
0.50		79	7.33 ± 0.66^{b}	$3.16 \pm 0.16^{a,b}$
0.75		71	5.66 ± 0.88^{d}	2.30 ± 0.20^{c}
1.00		65	$4.33 \pm 0.33^{d,e}$	$1.96 \pm 0.14^{d,e}$
	0.10	61	$4.66 \pm 0.33^{c,d,e}$	$1.83 \pm 0.16^{c,d}$
	0.25	75	$6.33 \pm 0.33^{b,c}$	2.26 ± 0.23^{c}
	0.50	93	9.66 ± 0.66^{a}	3.56 ± 0.23^{a}
	0.75	87	7.66 ± 0.66^{b}	2.83 ± 0.20^{b}
	1.00	81	$5.33 \pm 0.66^{c,d,e}$	$2.12 \pm 0.16^{d,e}$

Effect of full and half strength woody plant medium, supplemented with different concentrations of IBA on in vitro rooting in regenerated microshoots of *R. tetraphylla* after 4 weeks of incubation

11.3.5 Acclimatization and Hardening in Micropropagated Plants

The efficiency of micropropagation system for commercial purposes depends upon cost, production and survival rate of the plantlets under field conditions. The ex vivo transfer of tissue culture raised plants in critical phase where plantlets have to compete with new environment and the other biotic stresses (Hussain et al. 2019b; Naaz et al. 2018, 2019). Thus, all efforts are focused on ex vitro establishment of regenerants in order to reduce the stresses and mortality rate among regenerants. The micropropagated plants were acclimatized under culture room conditions successfully with 90% survival rate (Fig. 11.1d). After 4 weeks of proper hardening, the regenerated plants were planted in earthen pots filled with garden soil and thereafter pots were established in the green house where they grew well under full sun. All the tissue culture raised plants were morphologically alike with the mother plant (explant donor plant) and showed the similar growth pattern.

11.4 Conclusion

The model system developed will provide a novel concept for understanding the beneficial role of silicon in physiological and biochemical functions in plants at molecular level. The protocol established produced could be useful in enhancing and understanding the biomass production of pharmaceutically important alkaloids accumulated in *R. tetraphylla* cultures. Thus, for gaining more success in the plant tissue culture branch of plant biotechnology addition of Si with the culture medium may resolve and excel out the problems related to micropropagation.

Acknowledgements The author (Sheikh Altaf Hussain) is highly thankful to UGC (University Grants Commission—New Delhi) for the award of JRF/SRF under MANF scheme.

Conflict/Argument of Interests The author affirms that there is not any conflict of interest.

References

Ahmad N, Javed SB, Khan MI, Anis M (2013) Rapid plant regeneration and analysis of genetic fidelity in micropropagated plants of *Vitex trifolia*: an important medicinal plant. Acta Physiol Plant 35:2493–2500

Anis M, Ahmad N (2016) Plant tissue culture: propagation conservation and crop improvement. Springer, Singapore

Anonymous (2003) Wealth of India (raw materials), vol 8. CSIR, New Delhi, India, pp 398–399

Faisal M, Anis M (2009) Changes in photosynthetic activity, pigment composition, electrolyte leakage, lipid peroxidation, and antioxidant enzymes during ex vitro establishment of micropropagated *Rauvolfia tetraphylla* plantlets. Plant Cell Tissue Org Cult 99:125–132

Faisal M, Ahmed N, Anis M (2006) In vitro plant regeneration from alginate-encapsulated microcuttings of *Rauvolfia tetraphylla* L. Am Eurassian J Agricult Environ Sci 1:1–6

Hussain SA, Ahmad N, Anis M (2018a) Synergetic effect of TDZ and BA on minimizing the post-exposure effects on axillary shoot proliferation and assessment of genetic fidelity in *Rauvolfia tetraphylla* (L.). Rendiconti Lincei Fisiche e Naturali 29:109–115

Hussain SA, Ahmad N, Anis M, Alatar AA, Faisal M (2018b) Role of thidiazuron in modulation of shoot multiplication rate in micropropagation of *Rauvolfia* species. In: Ahmad N, Faisal M (eds) Thidiazuron: from urea derivative to plant growth regulator. Springer, Singapore, pp 429–438

Hussain SA, Ahmad M, Anis M, Alatar AA (2019a) Influence of *Meta* topolin on in vitro organogenesis in *Tecoma stans* L., assessment of genetic fidelity and phytochemical profiling of wild and regenerated plants. Plant Cell Tissue Org Cult 138:339–351

Hussain SA, Ahmad N, Anis M, Hakeem K (2019b) Development of an efficient micropropagation system for *Tecoma stans* [(L.) Juss. ex Kunth], using thidiazuron and effects on phytochemical constitution. In Vitro Cell Develop Biol Plant 55:442–453. https://doi.org/10.1007/s11627-019-10001-x

Hussain SA, Anis M, Alatar AA (2020) Efficient in vitro regeneration system for *Tecoma stans* L., using shoot tip and assessment of genetic fidelity among regenerants. Proc Natl Acad Sci India Sect B Biol Sci 90:171–178. https://doi.org/10.1007/s40011-019-01093-x

Islam MM, Ahmed M, Mahaldar D (2005) In vitro callus induction and plant regeneration in seed explants of rice (*Oryza Sativa* L.). Res J Agric Biol Sci 1:72–75

Jyothi T, Brijesh K, Hari Venkatesh KR (2012) Pharmacognostical evaluation of *Rauvolfia tetraphylla*. J Pharmaceut Sci Innovat 1:57–60

Khanam MN, Anis M (2018) Organogenesis and efficient in vitro plantlet regeneration from nodal segments of *Allamanda cathartica* L. using TDZ and ultrasound assisted extraction of quercetin. Plant Cell Tissue Org Cult 134:241–250

Ma JF (2004) Role of silicon in enhancing the resistance of plants to biotic and abiotic stresses. Soil Sci Plant Nutr 50:11–18

Mathe C, Mosolygó Á, Surányi G, Beke A, Demeter Z, Tóth VR, Beyer D, Mészáros I, Márta M (2012) Genotype and explant type dependent morphogenesis and silicon response of common reed (*Phragmites australis*) tissue cultures. Aquat Bot 97:57–63

Monney MAD, Amissah N, Blay E (2016) Influence of BA and IBA or NAA combinations on micropropagation of *Cryptolepis sanguinolenta*. Am J Plant Sci 7:572–580

Murashige T, Skoog F (1962) A revised medium for rapid growth and bioassays with tobacco tissue cultures. Physiol Plant 15:473–497

Naaz A, Hussain SA, Naz R, Anis M, Alatar AA (2018) Successful plant regeneration system via de novo organogenesis in *Syzygium cumini* (L.) Skeels: an important medicinal tree. Agrofor Syst 93:1285–1295

Naaz A, Hussain SA, Anis M, Alatar AA (2019) *Meta* topolin exposure improved micropropagation in *Syzgium cumini* and acclimatization to ex vitro conditions. Biol Plant 63:174–182

Perveen S, Javed SB, Anis M, Aref IM (2013) Rapid in vitro multiplication and ex vitro establishment of Caribbean copper plant (*Euphorbia cotinifolia* L.): an important medicinal shrub. Acta Physiol Plant 35:3391–3400

Prabagar S, Hodson MJ, Evans DE (2011) Silicon amelioration of aluminium toxicity and cell death in suspension cultures of Norway spruce (*Picea abies* L.). Environ Exp Bot 70:266–276

Sivanesan I, Jeong BR (2014) Silicon promotes adventitious shoot regeneration and enhances salinity tolerance of *Ajuga multiflora* Bunge by altering activity of antioxidant enzyme. Sci World J 2014:521703. https://doi.org/10.1155/2014/521703

Sivanesan I, Song JY, Hwang SJ, Jeong BR (2011) Micropropagation of *Cotoneaster wilsonii* Nakai—a rare endemic ornamental plant. Plant Cell Tissue Org Cult 105:55–63

TDZ-Induced Efficient Micropropagation from Juvenile Nodal Segment of *Syzygium cumini* (Skill): A Recalcitrant Tree

12

Afshan Naaz, Iram Siddique, and Altaf Ahmad

Abstract

An assessment was carried out to examine the TDZ (a potential cytokinin) efficacy in the induction of multiple shoot regeneration in *Syzygium cumini* from juvenile nodal (N) explants and cotyledonary node (CN) explants. TDZ in different concentration was used as a supplement to the MS basal media. Two factors viz. concentration and exposure duration to TDZ have multifaceted effect on multiple shoot induction and regeneration. Continuous presence of culture in TDZ inhibited shoot elongation and even leads to shoot fasciation. On transferring the TDZ exposed shoot clumps (4 weeks old) on hormone-free medium increased multiple shoot regeneration by threefold. Best TDZ concentration was recorded to be 5.0 μM in nodal explants while 2.5 in cotyledonary node explant. Maximum percentage response in CN explant was 75% with maximum mean shoot numbers (4.9 ± 0.67) and mean shoot length (2.90 ± 0.23 cm) while highest percent response was observed as 79% with maximum shoot numbers (4.17 ± 0.81) and maximum shoot length (2.87 ± 0.23 cm) in juvenile nodal (N) explants. The shoot clumps from CN (2.5 μM, TDZ) and juvenile nodal (5.0 μM, TDZ) when transferred onto secondary medium (hormone-free basal medium) produced mean shoot length (4.27 ± 0.34 cm) and (4.56 ± 0.35 cm) whereas mean shoot numbers noted to be (16.23 ± 0.21) and (18.34 ± 0.35), respectively. These shoot cultures are free from any discrepancy. Well-developed shootlet (3.0 cm) was successfully rooted in vitro in MS ½ and IBA with highest percent response in CN and juvenile nodal recorded to be 83% and 82% with mean root numbers (6.33 ± 0.10) and (5.99 ± 0.11) respectively and mean root length (7.19 ± 0.21 cm) and (6.79 ± 0.25 cm). The survival rate of transferred plant was nearly 70%.

A. Naaz (✉) · I. Siddique · A. Ahmad
Plant Biotechnology Laboratory, Department of Botany, Aligarh Muslim University, Aligarh, India

I. Siddique (ed.), *Propagation and Genetic Manipulation of Plants*,
https://doi.org/10.1007/978-981-15-7736-9_12

Keywords

Carry-over effect · Secondary medium · TDZ exposure · Cotyledonary · Juvenile nodal · Acclimatization · Exposure duration

12.1 Introduction

TDZ is highly potent plant growth regulator mimicking cytokinin-like activity (Mok et al. 1982) being 20 times more effective in breaking bud dormancy (Wang et al. 1986) including recalcitrant and woody trees where other cytokinins are ineffective (Huetteman and Preece 1993). Chemically it is phenyl urea derivatives, synthesized by German Schering Corporation for defoliation of *Gossypium hirsutum* (Suttle 1985). Morphogenesis has been described as a process that occurs in three distinct stages of development, namely (a) acquisition of competence, (b) induction or determination, and (c) differentiation (Christianson and Warnick 1983). The induction process on competent tissues leads to the formation of morphogenic cells, which become determined on the morphogenic pathway. The above separation is ultimately crucial in the scaling up and mechanization of micro-propagation systems. In plant tissue culture, it has been established as a potent plant growth regulant of a wide array of plant species including trees and medicinal plant species for the development of sustainable attainable morphogenetic system. The mechanism of TDZ action is partially related to the inhibition of cytokinin degradation by cytokinin oxidase, resulting in an increased level of endogenous cytokinin (Hare and Van Staden 1994), by easy translocation of auxin within the tissue (Murthy et al. 1998) accompanied by high acid–phosphate level (Mok et al. 1987). Induction of callogenesis, embryogenesis, and caulogenesis could all be achieved by augmenting TDZ with basal media followed by transferring the culture clumps into secondary medium of desirable choice. Shoot regeneration in TDZ exposed media is concentration-dependent. It is highly stable and biologically active at lower concentration (1–5 μM). Higher concentration of TDZ implicit many morphological defects like tissue deformity, improper intermodal growth, vitrification or fasciations, and so on, in a micropropagation system, significantly affecting height of newly formed shoot.

Syzygium cumini is an evergreen tropical tree belonging to family Myrtaceae. It is one of the most widely used medicinal plant of which all of the part are being used for therapeutic use or in many industries related to food and beverage, shipping, construction, wood, and so on, and is highly recommended for rejuvenation of forestry program. Pharmaceutical research and literatures across the world show that seeds, bark, leaf extract, and fruits of the plant have great therapeutic value and are being used in Indian subcontinent as an alternative medicine for curing of diabetes mellitus (Ravikumar et al. 2012) since ages as mentioned in *Charak samhita* (ancient medical book guidance). The steroids and saponins obtained from leaf sap help in the treatment of Alzheimer disease. The plant extract is beneficial in its action for killing of cancer cells (Swami et al. 2012) with reference to the antineoplastic activities. Owing to its high medicinal value and industrial demand economical cultivation of elite species of Indian blackberry is underway in many parts of Maharashtra particularly Vidarbha regions (Sirsat et al. 2019). There is mass-scale production of Jamun seedling of elite and economical species under government-sponsored entrepreneurship program that could be possible to generate additional revenue. However, it still lacked a proper protocol for economical and large-scale production. *S. cumini* suffers from very low seed viability and poor germination in its natural habitat (Dent 1948) and the seeds are very much prone to fungal infection (shelf life is hardly 3–4 days). Propagation through stem cuttings is an alternative, but not feasible for obtaining large quantity of planting materials. In this context, it is necessary to standardize a suitable protocol for clonal propagation of this species. Forest trees in general have proved to be difficult for mass propagation by tissue culture.

In this present communication, a reliable sustainable protocol for micropropagation from nodal and cotyledonary explants procured from 15 days old in vitro raised seedling has been presented. Very limited reports are there for the micropropagation of *S. cumini*. Given the fact that so far in vitro regeneration studies using TDZ has not been reported in *S. cumini*. The present study was aimed at working out a method of micropropagation using nodal and cotyledonary segments in response to different media and different concentrations of PGRs (TDZ, BA) for maximum plantlet production. Besides this, studies have been conducted to see the role of carryover effect on rooting acclimatization and antioxidant activity. Exposure duration of TDZ has also been taken into consideration as it had earlier been proved the damaging effect of extreme exposure on morphogenetic potential under in vitro conditions.

12.2 Materials and Methods

12.2.1 Establishment of Aseptic Culture

Shoot multiplication of *S. cumini* was evaluated by varying different parameters like type of basal media (MS and WPM), cytokinins (TDZ), and auxins (NAA and IBA) singly or in combinations during multiplication and rooting phase. In the first set of experiment, MS medium and WPM were used to identify the suitable basal media. The experimental materials (nodal explants and cotyledonary node (CN) from 15 days old seedling grown under in vitro condition in the culture room (as described by Naaz et al. 2019) were procured. The CN and nodal segments were cut (1.5 cm) from the in vitro raised seedlings under laminar flow following by inoculation into autoclaved flask (Borosil 200 mL, wide mouth) and directly inoculated into basal media with desirable hormonal concentration of TDZ (0.1–7.50 μM). MS (Murashige and Skoog 1962) and WPM (Lloyd and McCown 1981) basal media supplemented with 3% (w/v) sucrose and gelled with 0.8% (w/v) bacteriological grade agar (Qualigens, India) augmented with TDZ (0.1–7.5 μM) with the pH maintained to 5.8 using 1 N NaOH or HCl before autoclaving at 121 °C for 17 min. The explants (nodal) and CN were cultured in 50 mL flask (Borosil,

India) each containing 30 mL of medium. All the cultures were maintained at 24 ± 2 °C in a culture room illuminated by two cool white fluorescent lamps (40 W, Phillips India) with an irradiance of 50 μmol m^{-2} s^{-1} provided by cool white fluorescent tubes (40 W, Phillips) for 16 h/day. After 2, 3, and 4 weeks of incubation, the TDZ-conditioned explants were transferred into basal medium lacking TDZ. Both the explants (nodal and CN) were inoculated in MS or WPM. Since MS response was much better than in WPM; therefore, further experiments were carried out in MS basal medium. The frequency of responding explants, the number of regenerated shoots, and shoot length were recorded after 4 weeks of incubation in the TDZ-free medium. All cultures were subcultured onto the same fresh medium after 3–4 weeks.

12.2.2 Rooting and Acclimatization

The in vitro root induction was performed by transferring the individual microshoots on full strength MS and half-strength MS medium amended with root inducing hormones namely IBA in different concentrations (1.0, 2.5, 5.0, 7.5 μM). Data on percentage of rooting and mean number of roots per shoot were recorded after 3 weeks of transplantation. All culture conditions were the same as described for the shoot bud induction. The rooted shoots were washed gently under running tap water and transplanted into 10 cm diameter thermocol cups containing autoclaved soilrite (Keltech Energies Ltd., Bangalore, India) and covered with transparent plastic bags to ensure high humidity. They were further kept at low illumination with an irradiance of 50 μmol m^{-2} s^{-1} (cool white fluorescent lamp, 40 W) for 1 week and watered regularly with half-strength MS solution lacking both vitamins and PGRs over a period of 8–10 days. The plants were then transferred to pots containing normal garden soil and maintained in greenhouse conditions.

12.3 Results and Discussion

12.3.1 Effect of TDZ on Nodal Explants

Nodal explants grown on MS medium lacking PGR did not show any morphogenic response and failed to produce shoots even after 4 weeks of incubation. However, the medium supplemented with TDZ effectively induced bud break in nodal explants at all the concentrations tested (0.1, 0.5, 1.0, 2.5, 5.0, and 7.5 μM). Explants cultured on MS and WPM basal medium alone served as control and did not promote axillary bud induction leading to necrosis. However, explants (nodal and CN) both exhibited a better response of bud break and shoot induction in MS medium to that of WPM. The recorded data indicated that MS basal medium promoted better shoot multiplication than WPM which was in contrast to earlier report by Ramashree et al. (2007). However, this result was in consonant with earlier report by Naaz et al. (2019) where MS was supposedly better than WPM. MS basal medium supplemented with TDZ

Table 12.1 Effect of TDZ on multiple shoot induction from nodal explants derived from aseptic seedlings after 4 weeks of culture

TDZ (μM)	% Response	Mean number of shoots/explant	Mean shoot length (cm)
0.0	0.0	0.0	0.0
0.1	50.23	3.56 ± 0.45^{b}	2.05 ± 0.03^{c}
0.5	68.07	$3.67. \pm 0.22^{b}$	2.10 ± 0.23^{c}
1.0	75.05	$3.89 \pm 0.62^{a,b}$	2.60 ± 0.13^{b}
2.5	75.00	4.01 ± 0.81^{a}	2.70 ± 0.22^{b}
5.0	**79.00**	$\mathbf{4.17 \pm 0.81^{a}}$	$\mathbf{2.87 \pm 0.23^{a}}$
7.5	65.26	3.50 ± 0.22^{b}	2.20 ± 0.21^{c}

Values represent means ±SE. Means followed by the same letter within columns are not significantly different ($P = 0.05$) using Duncan's multiple range test

effectively induced bud break in nodal explants at all the concentrations tested (0.1, 0.5, 1.0, 2.5, 5.0, and 7.5 μM) within 2 weeks of incubation. Shoot bud proliferation was enhanced by increasing the concentration of TDZ up to an optimal level (5.0 μM) in the MS medium beyond which there is declination in all parameters taken into consideration. The highest percent frequency (79%), maximum number (4.17 ± 0.81) of shoots per explant with shoot length (2.87 ± 0.23 cm) was obtained on MS medium containing TDZ (5.0 μM), after 4 weeks of culture (Table 12.1; Fig. 12.1a). Higher concentrations (>5.0 μM) promote callusing and fasciations of shoots in some cultures. Also, a prolonged exposure to TDZ beyond 3 weeks resulted in fasciation, stunted growth, and hyperhydric shoots and leaves (Fig. 12.1b). After an initial exposure of TDZ for 4 weeks, the shoot clumps on transferring onto secondary medium (growth regulator free) devoid of TDZ enhance shoot multiplication markedly and the culture was more healthy and vigorous without any discrepancy (Fig. 12.1c). Culture exposed to TDZ (5.0 μM) generated maximum shoots (18.34 ± 0.35)/explant with maximum shoot length (4.56 ± 0.34 cm) (Fig. 12.2). However, basal medium (WPM) along with different TDZ concentrations showed a different response. The bud break was noted at concentration of 2.5 μM and beyond. The lower concentrations only showed swollenness in the explants that later got necrosis within 4 weeks. The TDZ result was not satisfactory.

The morphogenic responses of cotyledonary node (CN) explants on MS medium supplemented with different concentrations of TDZ (0.1, 0.5, 1.0, 2.5, 5.0, and 7.5 μM) were explored and the results are summarized in Table 12.2. All the concentrations of TDZ with MS facilitated shoot bud induction. Shoot regeneration frequencies and the number of shoots per explant declined with an increasing TDZ concentration (0.5–7.5 μM) as in the case of nodal explants. Beyond the optimum level (2.5 μM) and incubation period beyond 4 weeks the culture exhibited hyperhydricity and fasciation growth cessation failing to produce healthy shoots (Fig. 12.1c). The highest regeneration frequency was recorded to be 75% with maximum mean number of shoots per explant (4.9 ± 0.67) and mean shoot length (2.90 ± 0.23 cm) (Table 12.2). Subculturing to the same medium was futile for

Fig. 12.1 (**a**) Induction of shoots on MS+ TDZ 2.5 (μM) in nodal explant after 4 weeks, (**b**) fasciated shoot and hyperhydricity in TDZ-mediated MS medium beyond 4 weeks of culture, (**c**) shoot multiplication and elongation on hormone-free medium after 4 weeks in CN explant, (**d**) rooting of microshoots in MS½ + IBA (**e**) acclimatized plant after 6 weeks of hardening

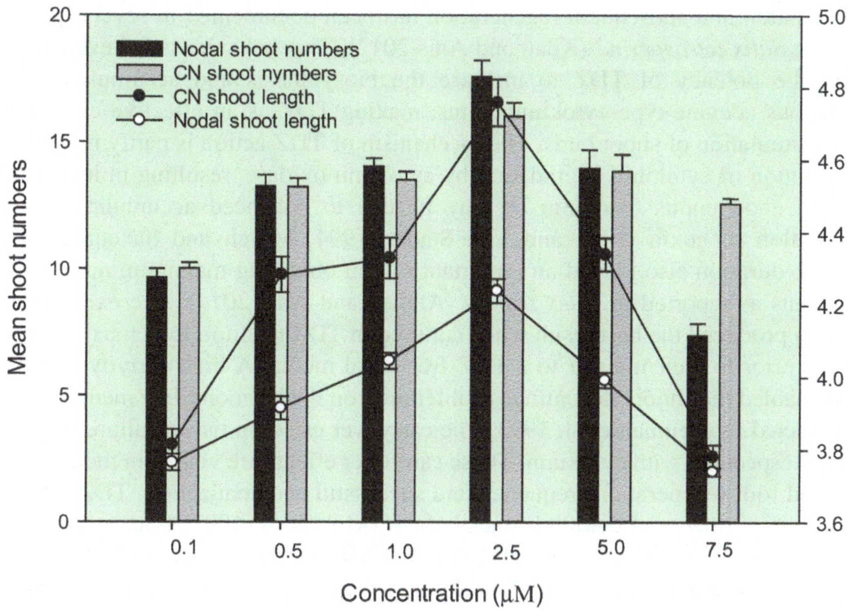

Fig. 12.2 Effect of TDZ on shoot proliferation from CN and nodal explants after 4 weeks of transfer to MS basal media (TDZ-free media). Bars represent the means ± SE. Bars denoted by the same letter within response variables are not significantly different ($P = 0.05$) using DMRT. Evaluation was made after 8 weeks of culture

Table 12.2 Effect of TDZ on multiple shoot induction from cotyledonary node (CN) explants derived from aseptic seedlings after 4 weeks of culture

TDZ (μM)	% Response	Mean number of shoots/explant	Mean shoot length (cm)
0.0	0.0	0.0 ± 0.0^{e}	0.0 ± 0.0^{d}
0.1	50.23	2.67 ± 0.34^{d}	2.05 ± 0.21^{c}
0.5	60.87	3.90 ± 0.45^{b}	2.10 ± 0.23^{c}
1.0	71.56	4.50 ± 0.22^{a}	2.60 ± 0.23^{b}
2.5	**75.04**	$\mathbf{4.90 \pm 0.67^{a}}$	$\mathbf{2.90 \pm 0.23^{a}}$
5.0	71.34	3.20 ± 0.81^{c}	2.61 ± 0.22^{b}
7.5	55.26	3.10 ± 0.22^{c}	2.20 ± 0.21^{c}

Values represent means ±SE. Means followed by the same letter within columns are not significantly different ($P = 0.05$) using Duncan's multiple range test (*DMRT*)

recovering growth in the cultures; therefore, the cultures were transferred to MS basal media devoid of TDZ. This resulted in the resurgence of shoot differentiation in the cultures. Maximum growth of shoot numbers and shoot length was recorded in cultures which were initially cultured on MS supplemented with 2.5 μM TDZ fetching about 16.23 shoots per explant with an average shoot length of 4.27 ± 0.34 cm after 8 weeks of culture (Fig. 12.2). These cultures were healthy and without any discrepancy (Fig. 12.1c). The effectiveness of TDZ in shoots bud differentiation and subsequent regeneration has been documented in several plants including *Salix tetrasperma* (Khan and Anis 2012). This behavior is believed to be due to the potency of TDZ to increase the biosynthesis and accumulation of endogenous adenine-type cytokinins, thus, making TDZ as an effective cytokinin for the stimulation of shoot buds. The mechanism of TDZ action is partly related to the inhibition of cytokinin degradation by cytokinin oxidase, resulting in increased levels of endogenous cytokinin or may be due to enhanced accumulation and translocation of auxin (Hare and Van Staden 1994; Murch and Saxena 2001). Exposure duration also played an important role in obtaining maximum number of regenerants as reported in *Vitex trifolia* (Ahmed and Anis 2014). The explants of *S. cumini* produced the best result when cultured in TDZ medium for a maximum of 4 weeks prior to their transfer to a TDZ free basal media. A high carryover effect which enabled the shoots to continue proliferation on the hormone-free medium was also depicted by Neuman et al. 1993. The carryover effect in tissue culture is quite common especially with cytokinin. These carryover effects are visible in the form of shoot and root regeneration frequency and successful acclimatization. TDZ affects pathway of purines and cytokinin metabolism (Mok et al. 1982; Laloue and Fox 1989). It involves in modification of cell membrane, energy levels, nutrient uptake, and nutrient assimilation. Our result corroborates findings by Faisal et al. (2005).

12.3.2 Effect of Exposure Duration of TDZ

TDZ exposed culture showed time-dependency effect. On increasing the exposure duration, the highest concentration (10.0 μM) did the maximum damage to the tissues while lower concentration (less than 5.0 μM) could sustain the tissue lines even after 8 weeks exposure (data not shown). A 4 week exposure to the optimal TDZ concentration (5.0 μM) was preferred as this culture regime amounted to maximum shoot yield per explant (14.91 $\pm$ 1.10) on transfer to secondary medium (devoid of growth regulator) after 4 weeks (Fig. 12.3). An exposure duration exceeding 4 weeks elicited lower shoot induction afterwards. Up to maximum exposure duration of 6 weeks, nodal explants expressed regeneration capability to produce new shoots. The exposed tissues turned brown and necrotic after 6 weeks and no further regeneration was observed on changing the media. Moreover, continuous exposure of culture (more than 4 weeks) on TDZ led to abnormal shoot morphology, fasciation, and distortion (Fig. 12.1b) while on the other hand, shorter exposure time (2 weeks) could not elicit proper bud break and thus was incapable to produce sufficient and desired results. Our result is in congruence with Khan and Anis (2012) (Fig. 12.2).

12.3.3 Effect of Subculturing

TDZ containing media could not affect multiplication and elongation of induced shoots. Therefore these TDZ exposed shoots were further transferred in secondary

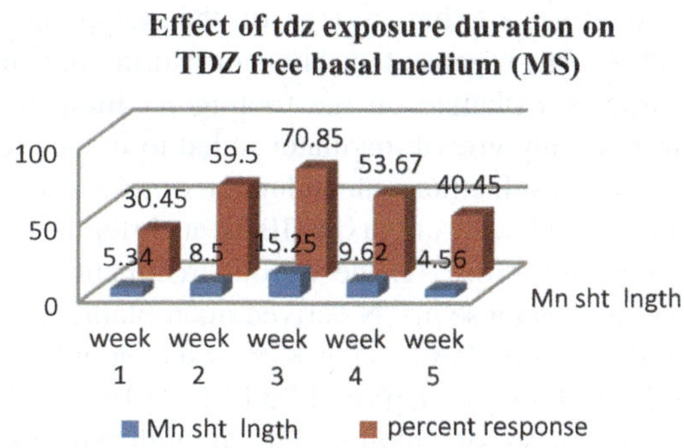

Fig. 12.3 Effect of TDZ-exposure duration on TDZ free medium (MS) on nodal explants

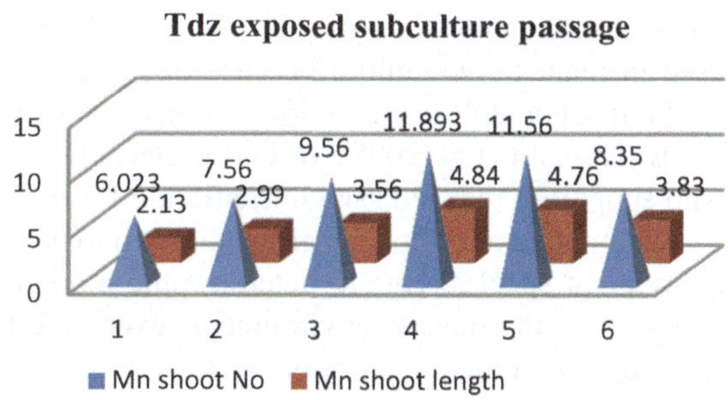

Fig. 12.4 Effect of TDZ exposure duration of nodal culture on TDZ-free basal medium (MS)

medium for their further growth and proliferation. In the present study, the optimal TDZ (2.50 and 5.0 μM) exposed explants (CN and nodal) carry a high potential of multiplication on a basal media devoid of any growth regulators. This transfer in secondary medium after TDZ exposure greatly reduced the morphological and physiological disorders and increased the regeneration potential of the explants. Our result corroborates earlier findings by Khan and Anis (2012) and Dewir et al. (2018). In the first four subculturing passages, the rate of multiplication of shoot buds was found to increase and beyond which decline in multiplication rate was noticed. The well-developed shoots were regularly removed and transferred to rooting medium (Fig. 12.4).

12.3.4 In Vitro Rooting and Acclimatization

Rooting of in vitro regenerated shoots and transplantation of the plantlets to the field is the most important, crucial, and essential step (Pospisilova et al. 1999), but difficult task in tissue culture of woody trees. In the present study, optimization of rooting response of micro shoots of *S. cumini* was achieved using different auxins at varying concentration. Translocation of IBA and IAA has a very critical role in rooting response in woody tree species due to complex mechanism involved wherein carryover effect of the previously induced cytokinins during multiplication phase also effect rooting system of the newly developed plantlets. (Rashotte et al. 2003). The regenerated shoots measuring 3.4 cm were excised from the regenerated tissue and planted on the rooting medium. MS (full and half strength) medium without any growth regulator failed to induce root formation even after 4 weeks of transfer. To facilitate the induction of root, MS basal and half strength medium were augmented with an auxin (IBA) at different concentration (1.0, 2.5, 5.0, 7.5 μM) with 3% sucrose. Of the various treatments evaluated for in vitro rooting, the best rooting response in CN derived microplants was observed in half-strength MS along with 5.0 μM IBA, giving 83% response with average number of 6.0 roots with a length of 7.0 cm approx. (Table 12.3; Fig. 12.1d). Initiation of roots was observed after 7 days of culture. Half-strength MS showed promotory effect on rooting response in comparison with full strength MS. Whereas microshoots derived from nodal explants showed percentage response of 82% and highest shoot numbers to be six with maximum shoot length 6.79 ± 0.25 cm (Table 12.3). IBA has often been used in plant tissue culture because of its easy translocation (Davies et al. 1982; Rashottte et al. 2003). Our reports are in consonance with earlier reports in *Sterculia urens* (Hussain et al. 2008), *Balanites aegeptiaca*, and *S. cumini* (Naaz et al. 2014) who supported reduced salt concentration for better rhizogenesis.

Acclimatization of plants grown in vitro to natural conditions is a crucial step for many plant species. Due to heterotrophic mode of nutrition, lack of adaptation or exposure to the outside environment (excessive temperature and humidity, atmospheric CO_2, mineral nutrition, pH, and texture of soil) affects the regenerated plantlets. Maximum acclimatization percentage was noted to be 70% in soilrite (Fig. 12.1e). These results are in congruent with the findings of Anis et al. (2010) and Perveen and Anis (2015), where regenerants showed highest survival in soilrite. The plants appeared morphologically healthy with normal leaf, shape, and growth patterns.

12.4 Conclusion

The present communication describes the effectiveness of TDZ as potent cytokinin for bud breaking, shoot multiplication, healthy culture, and acclimatization in *S. cumini* (L.). The TDZ-exposed cultures could induce better shoot regeneration on being transferred to hormone-free basal medium thereby protecting the juvenile shootlet from carryover effect of the hormones. There are also better survival and adaptation of the juvenile plants to the harsh weather and high sunlight and humidity outside. The protocols could be efficiently used for large-scale regeneration of other woody and recalcitrant species to produce disease-free clones without any seasonal variation. Ex vitro acclimatization is the final step that owes a major influence on any micropropagation system. In many cases, this is a major bottleneck and needs to be approached carefully, considering carryover effects and stress tolerance from the in vitro phase toward the external environmental condition. Numerous protocols have exemplified that TDZ can successfully replace the commonly used CKs (BA, KN, Zeatin, 2ip) in many plant tissue culture protocols and can even produce better

Table 12.3 Effect of IBA on root regeneration from nodal explants on ½ MS medium after 4 weeks of culture

Treatments (μM)	IBA (μM)	Response (%)		Mean root number		Mean root length (cm)	
		Nodal (N)	CN	Nodal (N)	CN	Nodal (N)	CN
½ MS	0.0	0.0 ± 0.0^{e}	0.0 ± 0.0^{e}	0.0 ± 0.0^{e}	0.0 ± 0.0^{e}	0.0 ± 0.0^{e}	0.0 ± 0.0^{e}
½ MS	1.0	$75.05 \pm 0.67^{d,e}$	$73.09 \pm 0.43^{d,e}$	2.50 ± 0.10^{d}	2.67 ± 0.11^{d}	5.20 ± 0.17^{b}	4.80 ± 0.19^{b}
½ MS	2.5	80.02 ± 0.41^{b}	80.12 ± 0.23^{b}	4.10 ± 0.15^{b}	4.04 ± 0.25^{b}	6.03 ± 0.18^{c}	5.63 ± 0.28^{c}
½ MS	**5.0**	**83.05 ± 0.18^{a}**	**82.75 ± 0.23^{a}**	**6.33 ± 0.10^{a}**	**5.99 ± 0.11^{a}**	**7.19 ± 0.21^{a}**	**6.79 ± 0.25^{a}**
½ MS	7.5	78.03 ± 0.54^{c}	77.73 ± 0.22^{c}	2.45 ± 0.15^{c}	2.56 ± 0.25^{c}	$6.05 \pm 0.23^{b,c}$	$6.01 \pm 0.13^{b,c}$

Values represent means ±SE. Means followed by the same letter within columns are not significantly different ($P = 0.05$) using Duncan's multiple range test (*DMRT*)

results where the routinely used CKs are not much of success and have inhabitant effect at various steps in woody tree species. The effectiveness of TDZ lies in the fact that they can play pivotal roles on some physiological response or disorders. Beside shoot multiplication carryover effect should be taken into consideration as accumulation of hormones in the tissue culture shootlets inducing toxicity, which inhibits proper growth of cultures as well as rooting and acclimatization process as a whole. Such an approach will elucidate us solving many constraints associated with physiological and developmental problems in vitro of woody tree species.

References

Ahmed R, Anis M (2014) Role of TDZ in the quick regeneration of multiple shoots from nodal explant of *Vitex trifolia* L.—an important medicinal plant. Appl Biochem Biotechnol 168:957–966

Christianson ML, Warnick DA (1983) Competence and determination in the process of in vitro shoot organogenesis. Dev Biol 95:288–293

Davies FT, Lazarte JE Jr, Joiner JN (1982) Initiation and development of roots in juvenile and mature leaf bud cuttings of *Ficus pumila* L. Am J Bot 69:804–811

Dent TV (1948) Seed storage with particular reference to the storage of seeds of Indian Forest plants. Indian Forest Research (N.S.). Silviculture 7:1–124

Dewir YH, Nurmansyah NY, Teixeira da Silva JA (2018) Thidiazuron-induced abnormalities in plant tissue cultures. Plant Cell Rep 7:1451–1470. https://doi.org/10.1007/s00299-018-2326-1

Faisal M, Ahmad N, Anis M (2005) Shoot multiplication in *Rauvolfia tetraphylla* L. using thidiazuron. Plant Cell Tissue Organ Cult 80:187–190

Hare PD, Van Staden J (1994) Inhibitory effect of TDZ on the activity of cytokinin oxidase isolated from soybean callus. Plant Cell Physiol 35:1121–1125

Huetteman CA, Preece JE (1993) Thidiazuron: a potent cytokinin for woody plant tissue culture. Plant Cell Tissue Org Cult 33:105–119

Hussain TM, Chanrashekher T, Gopal GR (2008) Micropropagation of *Sterculia urens Roxb* an endangered tree species from intact seedling. Afr J Biotechnol 7:95–101

Khan I, Anis M (2012) Modulation of in vitro morphogenesis in nodal segments of *Salix tetrasperma* Roxb. through the use of TDZ, different media types and culture regimes. Agrofor Syst 86:95–103

Laloue M, Fox JE (1989) Cytokinin oxidase from wheat. Plant Physiol 90:899–906

Lloyd G, McCown B (1981) Commercially feasible micropropagation of mountain laurel, *Kalmea latifolia*, by the use of shoot tip culture. Int Plant Propagat Soc 30:421–427

Mok MC, Mok DWS, Armstong DJ, Shudo K, Isogai Y, Okamoto T (1982) Cytokinin activity of N-phenyl-N'-1, 2, 3-thidiazol-5-ylurea (thidizuron). Phytochemistry 21:1509–1511

Mok MC, Mok D, Turner J, Mujer C (1987) Biological and biochemical effects of cytokinin-active phenylurea derivatives in tissue culture systems. In: chemical regulation in tissue culture: an overview. Hortic Sci 22:1194–1197

Murashige T, Skoog F (1962) A revised medium for rapid growth and bioassays with tobacco tissue cultures. Physiol Plant 15:473–497

Murch SJ, Saxena PK (2001) Molecular fate of thidiazuron and its effect on auxin transport in hypocotyls tissues of *Pelargonium hortorum* bailey. Plant Growth Regul 35:269–275

Murthy BNS, Murch SJ, Saxena P (1998) Thidiazuron: a potent regulator of in vitro plant morphogenesis. In Vitro Cell Develop Biol Plant 34:267–275

Naaz A, Altaf S, Naz R, Anis M, Alatar AR (2019) Successful plant regeneration system via de novo organogenesis in *Syzygium cumini* (L.) Skeels: an important medicinal tree. Agrofor Syst 93:1285–1295. https://doi.org/10.1007/s10457-018-0236-4

Naaz A, Shahzad A and Anis M (2014) Effect of Adenine Sulphate Interaction on Growth and Development of Shoot Regeneration and Inhibition of Shoot Tip Necrosis Under In Vitro Condition in Adult Syzygium cumini L.—a Multipurpose Tree. Appl Biochem Biotechnol.173:90–102

Neuman MC, Preece JE, Van Sambreek JW, Gaffney GR (1993) Somatic embyogenesis and callus production from cotyledon explants of black walnut (*Juglans nigra* L.). Plant Cell Tissue Org Cult 32:9–18

Perveen S, Anis M (2015) Physiological and biochemical parameters influencing ex vitro establishment of the in vitro regenerants of Albizia lebbeck (L.) Benth.: an important soil reclaiming plantation tree. Agrofor Syst 89:721–733

Pospisilova J, Ticha I, Kadlecek P, Haisel D, Plzakova S (1999) Acclimatization of micropropagated plants to ex vitro conditions. Biol Plant 42:481–497

Ramashree AB, Varghese TT, Raghu AV, Nabeese NN (2007) Micropropagation of *Syzygium cumini* skeels. A multipurpose tree. Res J Bot 2:208–213

Rashotte AM, Poupart J, Waddell CS, Muday GK (2003) Transport of the two natural auxins, indole-3-butyric acid and indole-3- acetic acid, in *Arabidopsis*. Plant Physiol 133:761–772

Ravikumar BVV, Patnaik AK, Kumar RS, Rao NN (2012) A RP-HPLC method development and validation for the estimation of montelukast sodium in bulk and pharmaceutical dosage forms. Int Res J Pharm 3:150–153

Sirsat M, Darasing R, Rathod Santosh Gahukar J, Akhare AA, Gawande VL, Maske NC, Munde BS, Padole DA, Rathod AV (2019) In vitro shoot proliferation from nodal segments of Indian blackberry (*Syzygium cumini* L.) Vividha. Int J Curr Microbiol App Sci 8:2464–2469

Suttle JC (1985) Involvement of ethylene in the action of the cotton defoliant Thidiazuron. Plant Physiol 78:272–276. https://doi.org/10.1104/pp.78.2.272

Swami S, Thakor N, Patil M and Haldankar P (2012) Jamun (Syzygium cumini (L.): A Review of Its Food and Medicinal Uses. Food and Nutrition Sciences, Vol. 3 (8) 1100-1117. doi: 10.4236/fns.2012.38146.

Wang SY, Steffens GL, Faust M (1986) Breaking bud dormancy in apple with plant bioregulator, thidiazuron. Phytochemistry 25:311–317

The manufacturer's authorised representative in the EU is Springer Nature Customer Service Centre GmbH, Europaplatz 3, 69115 Heidelberg, Germany. If you have any concerns regarding our products, please contact ProductSafety@springernature.com

Printed and bound by CPI Group (UK) Ltd, Croydon, CR0 4YY
30/09/2026
02225306-0015